INVENTORS and INVENTIONS

Volume 3

Fulton, Robert –
Kettering, Charles

Marshall Cavendish Reference
New York

Marshall Cavendish
99 White Plains Road
Tarrytown, New York 10591-9001

www.marshallcavendish.com

© 2008 Marshall Cavendish Corporation

All rights reserved, including the right of reproduction in whole or in any part in any form. No part of this book may be reproduced or utilized in any form or by any means electronic or mechanical, including photocopying, recording, or by any information storage and retrieval system, without prior written permission of the publisher and copyright owner.

Library of Congress Cataloging-in-Publication Data

Inventors and inventions.
    p. cm.
Includes bibliographical references and index.
ISBN 978-0-7614-7761-7 (set)—ISBN 978-0-7614-7763-1 (v. 1)— ISBN 978-0-7614-7764-8 (v. 2)— ISBN 978-0-7614-7766-2 (v. 3)—ISBN 978-0-7614-7767-9 (v. 4)—ISBN 978-0-7614-7768-6 (v. 5) 1. Inventions—History. 2. Technology—History. 3. Inventors—History.
I. Marshall Cavendish Corporation. II. Title.

T15.I62 2007
609--dc22

2007060868

Printed in Malaysia

11 10 09 08 07   1 2 3 4 5

Consulting Editor: Doris Simonis, Kent State University

Contributors: Richard Beatty; Jonathan Dore; Laura Lambert; Paul Schellinger; Mary Sisson; Gwendolyn Wells; Chris Woodford

MARSHALL CAVENDISH
Editor: Evelyn Ngeow
Publisher: Paul Bernabeo
Production Manager: Michael Esposito

MTM PUBLISHING
President: Valerie Tomaselli
Executive Editor: Hilary W. Poole
Editorial Coordinator: Tim Anderson
Editorial Assistants: Zachary Gajewski, Shalini Tripathi
Illustrator: Richard Garratt
Copyeditor: Carole Campbell

Design: Patrice Sheridan
Indexer: AEIOU, Inc.

Photographic Credits on page 960.

VOLUME 3

Fulton, Robert 645
  (Canal, submarine, and steamboat designs)
Gadgil, Ashok 657
  (Ultraviolet water purification system)
Galilei, Galileo 663
  (Telescope)
Getting, Ivan I. 675
  (Global Positioning System)
Gillette, King C. 683
  (Safety razor)
Ginsburg, Charles 691
  (Videotape recorder)
Goddard, Robert H. 699
  (Rockets)
Goldmark, Peter 711
  (Color television, long-playing record)
Goodyear, Charles 719
  (Vulcanized rubber)
Gourdine, Meredith 729
  (Device for purifying air)
Gutenberg, Johannes 737
  (Modern printing)
Hall, Charles 749
  (Aluminum manufacturing process)
Hall, Lloyd A. 757
  (Food preservation methods)
Hargreaves, James 763
  (Spinning jenny)
Harrison, John 773
  (Chronometer)
Health and Medicine 779
Henry, Beulah 791
  (Household devices)
History of Invention 797
Honda, Soichiro 805
  (Small motorcycle engine)
Hooke, Robert 815
  (Various scientific instruments)
Hopper, Grace Murray 821
  (Computer language compiler)
Household Inventions 827
Howe, Elias 837
  (Sewing machine)
Hunt, Walter 845
  (Safety pin)
Huygens, Christian 849
  (Telescope, pendulum clock)

Invention and Innovation 859
Jacquard, Joseph-Marie 865
  (*Jacquard loom*)
Jenner, Edward 873
  (*Smallpox vaccine*)
Jobs, Steve, and Steve Wozniak 883
  (*Apple I and Apple II computers*)
Johnson, Lonnie 893
  (*Super Soaker*)
Jones, Amanda 899
  (*Vacuum canning*)
Julian, Percy Lavon 905
  (*Glaucoma treatment, method of synthesizing cortisone*)

Kamen, Dean 913
  (*AutoSyringe, Segway*)
Kapany, Narinder 923
  (*Fiber optics*)
Keichline, Anna 931
  (*K brick*)
Kellogg, Will Keith 937
  (*Cornflakes*)
Kettering, Charles 943
  (*Self-starter for automobiles*)

Index 953
Photographic Credits 960

ROBERT FULTON

Civil and marine engineer and inventor
1765–1815

The 19th century was the age of the steam engine; the use of steam to power boats was a major development in trade, transportation, and technology. As the first person to design a commercially successful steam-powered watercraft, Robert Fulton helped inaugurate the era of fast and reliable travel by sea and river.

EARLY YEARS

Robert Fulton was born in Lancaster, Pennsylvania, on November 14, 1765. His father was a farmer who died when Robert was three years old. Fulton grew up wanting to be a painter; in 1782 he moved to Philadelphia, where he earned a modest but successful living as a painter of miniature portraits. At the age of 21 he went to London, where he began studies with the American-born painter Benjamin West (see box, Fulton the Artist).

After some years in London he began to tour the country, studying works of art in the country houses of the aristocracy and painting the owners' portraits. In Devon he met and befriended the earl of Stanhope, an amateur inventor who was interested in applying steam power to boats; and the duke of Bridgewater, who was a pioneer of canal-building in Britain. Through them he became interested in both boat-building and civil engineering projects. When he visited the manufacturing center of Birmingham, Fulton began to see what an impact a canal network could have in distributing goods.

CANAL PROPOSALS

Canals, however, presented problems. When a canal went through hilly country, changes in ground level could be only overcome by building expensive tunnels and aqueducts to maintain an even water level or by installing a series of locks—gated sections of the canal—to change the water level gradually. A boat would enter through one gate, and once the gates closed, water would be pumped into or out of the enclosed section, raising or lowering the water level. The boat would then exit by the opposite gate onto the next section of canal at the new water level. Combining several

Painting of Robert Fulton from around 1800 (artist unknown).

Fulton's drawing The Double Inclined Plane from 1796, shows the mechanism for moving boats along different levels of a canal using pulleys and metal tracks.

of these locks in sequence—making, in effect, a water "staircase"—allowed large changes of water level to be made. However, the process of filling or emptying each lock is very slow, and a boat might need hours to pass through a series of locks.

Fulton's solution was a variant of an already established idea: the inclined plane. This was a straight railroad built on a slope connecting two sections of canal at different levels. Cargo was put onto carts that were hauled up or down the track, then reloaded onto another barge on the other canal. This could be done in a small fraction of the time taken to go through a staircase of locks.

Fulton designed a system in which a large tub, suspended over a pit, was filled with water from the upper canal. As it filled and sank lower into the pit, the other end of the pulley system to which it was attached would pull a cart up the slope on one set of tracks, while a similar cart would be lowered down the slope on the set next to it. In 1794 Fulton received a British patent for this idea, and two years later he published *A Treatise on the Improvement of Canal Navigation*, in which he set out his proposals at length; it was the only major publication of his career. Fulton sent copies to U.S. president George Washington and other prominent members of the newborn republic, as he felt the ideas were well suited to the geography of the eastern United States.

THE SUBMARINE AND THE TORPEDO

Fulton was sympathetic to the democratic ideals of the French Revolution of 1789. Since 1793 Britain had been at war with France; American ships that attempted to trade at French ports had sometimes been seized by Royal Navy ships that were blockading the French coast. In 1797 Fulton went to Paris and offered to build the French government a weapon against the blockade.

Fulton's idea was a combination of weapons: a submarine and a mine. The submarine, which Fulton named the *Nautilus*, consisted of a copper skin on an iron frame and was the first "submarine boat" to have the cigar shape characteristic of almost all modern submarines. The mine, which

Undated lithograph of Fulton (left) meeting with Napoleon, the French emperor.

was towed on a rope behind the submerged submarine, consisted of a copper vessel filled with gunpowder that was triggered by contact with a solid object. Fulton called the mine a *torpedo*, after a type of fish that is capable of delivering an electric shock to anything that touches it. This was the first use of the term to describe an explosive device, though it was not until 1860 that the word came to be applied to the self-propelled underwater missile that we now call a torpedo.

Experiments in 1801 in the harbor at Brest, in the northwest of France, resulted in the sinking of a small test ship; Napoleon then gave Fulton permission to attack blockading vessels. Although he and his crew spent many weeks searching for a suitable target, the *Nautilus* was not fast enough to overtake any surface ship under sail, no attack took place, and the French government lost interest. In 1804 Fulton traveled to London, hoping to interest the Royal Navy, whose ships he had recently been trying to sink.

The prime minister, William Pitt, encouraged further developments, signing a contract with Fulton that paid him a large salary of £2,400 a year—equivalent to around $180,000 in modern currency—in addition to half the value of any enemy shipping sunk (ships were assigned a standard value according to their size and firepower, and the prize could be claimed if they were captured or destroyed). However, Pitt died in 1806, and the navy could see no way of overcoming the problem of inadequate speed that had dogged the design from the start. After the test run in Brest harbor, no ship was ever sunk by the *Nautilus* and its torpedo.

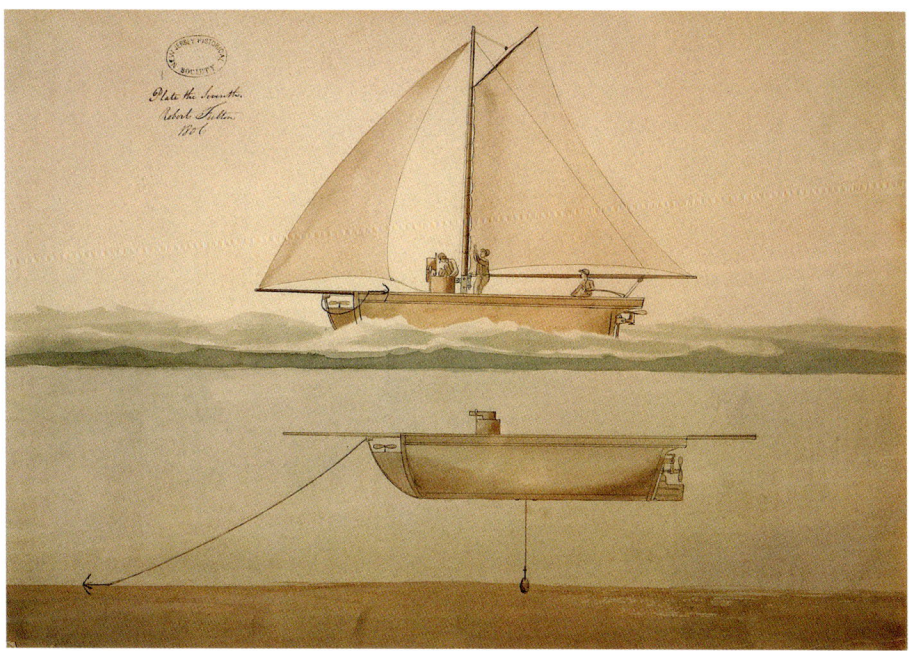

A drawing made by Fulton in 1806 of his submarine.

STEAMBOAT EXPERIMENTS

Meanwhile, Fulton was pursuing his interest in another new development: the use of steam engines to power surface ships. In 1801 Robert R. Livingston, one of the drafters of the Declaration of Independence, was appointed the U.S. minister (ambassador) to France. Livingston, who was also interested in steamboats, moved to Paris to carry out the responsibilities of his new position. Fulton was also then living in that city, and, in 1802, the two men signed a partnership document in which they agreed to try to build a workable steamboat that could be run at a profit on the Hudson River between New York City and Albany—thus combining Fulton's inventive and engineering skills with Livingston's high-level political and financial connections.

A letter Fulton wrote to the earl of Stanhope in 1793 shows that he had been experimenting with the best method of propulsion for a steamboat as early as 1788 (see box, Who Invented the Steamboat). Fulton conducted a series of experiments using small-scale models; his first idea was to mount a vertical oar at the back of a ship that would move back and forth, "to imitate the spring in the tail of a salmon." Fulton then thought of paddles, mounting a wheel with two paddles on each side of the ship; this design proved to be five times more efficient than the oar-propelled design. Further experimentation led Fulton to mount three paddles on each wheel, which made a craft move more smoothly through the water; more paddles made the wheels, mounted low in the water, less efficient. The later use of wheels with multiple paddles became feasible only when most of the wheel was lifted clear of the water, so that only a small number of paddles were submerged at one time.

A model (scale 1:27) of Fulton's first experimental steamboat.

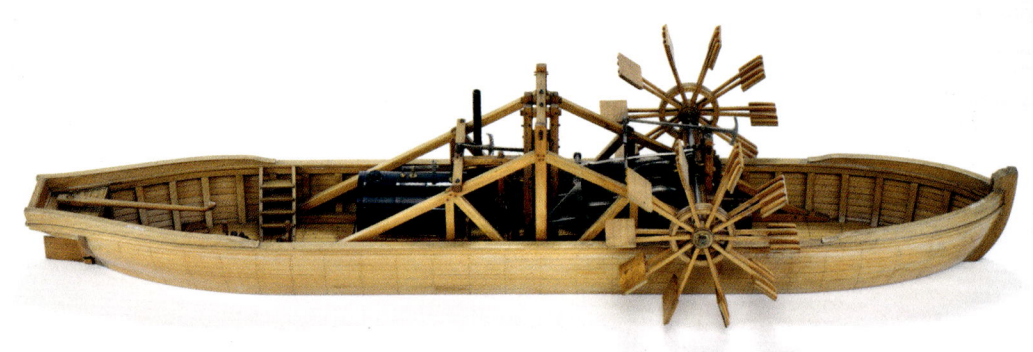

Who Invented the Steamboat?

In the 1600s, even before a steam engine that could do useful work had been built, people had begun to wonder how steam could be used to power ships. Once James Watt (1736–1819) had developed a steam engine powerful enough to drive a ship through the resistance of water, many people made the attempt to design and build a steam-powered vessel. In France, Claude Jouffroy d'Abbans (1751–1832) built a steam-powered paddleboat that he ran on the Doubs River in southern France in 1776. It failed shortly after setting out, but in 1783 an improved version ran successfully on the River Saône at Lyon. As d'Abbans ran out of money and the French government had no interest in the boat's further development, the steam-powered craft reached a dead end.

In the United States, John Fitch (1743–1798) built a steamboat that made a successful trial on the Delaware River in 1787. In 1790 he built another, propelled by ranks of oars like a mechanical Roman galley, which ran a successful passenger service on that river between Philadelphia and Trenton for several months. He, too, ran out of money. In Scotland, William Symington (1763–1831) built a steam engine into one of Patrick Miller's double-hulled boats, powering paddle wheels between the hulls in the manner of Fulton's final designs. It was tested successfully in October 1788, though a larger boat of similar design ran into difficulties in December 1789 on the Forth and Clyde Canal when the paddles began to break up under the power of the engine. Miller lost interest and cut off funding for the project.

In 1800, however, Symington found a new sponsor, Lord Dundas, who funded a new experimental canal boat, which ran trials on the Carron River the next year. Symington was then able to design and build a boat in which hull and engine were properly integrated. On March 28, 1803, the *Charlotte Dundas* pulled two barges weighing a total of 70 tons (64 metric tons) 18.5 miles (30 km) along the Forth and Clyde Canal in 9 hours 15 minutes. The canal owners still worried about eroding the canal banks, however, and the cancellation of an order for eight similar boats on another canal because of its owner's death spelled the end of the project. Fulton's genius in 1807 was to build a boat for a popular passenger route, ensuring the project's financial success and independence.

In 1802, Fulton conducted further experiments that confirmed his preference for side-mounted paddle wheels. With a borrowed steam engine, he constructed a boat that was about 74.5 feet long by 8.5 feet wide (22.7 by 2.6 m). In early 1803, before the boat was finished, it sank under the weight of its own engine. Fulton raised and rebuilt it; on August 9, 1803, he gave a demonstration in which the boat sailed up and down the Seine in Paris, making two trips upriver and two downriver at an average speed of 3.25 miles per hour (5.2 km/h). It pulled two smaller boats, in one of which sat several important members of the government (although Napoleon himself did not attend, as legend later suggested). Reaction was favorable, and the experiment was reported enthusiastically in the French press.

THE NORTH RIVER STEAMBOAT

Once he knew he could build a functioning steamboat, Fulton decided to return to the United States to put the plan for the New York–Albany run into operation, though he could not do so immediately, as his work on the *Nautilus* and its torpedo were to keep him busy for another three years. In December 1806, Fulton finally returned to his native land after an absence of more than 20 years.

He turned his attention immediately to the construction of the Hudson River steamboat. Notes he had written in 1802 show that he understood the principle that as boats grew larger, they required less energy per ton—or per passenger—to drive them (because a vessel's resistance to the water, and thus the energy needed to drive it, increases according to the square of its dimensions, but its volume increases by the cube of those dimensions). Fulton decided that his boat would have to be quite narrow and flat-bottomed to minimize its resistance, and long to maximize the number of people and weight of cargo it could carry. In the end, it was 150 feet (45 m) long, 13 feet (4 m) wide, and its bottom was only 2 feet 4 inches (71 cm) below the waterline. For a riverboat, it was much more important to be able to move in shallow water than to have the deeper hull that gives stability to oceangoing ships.

Although this boat is famous as the *Clermont*, Fulton always called it simply the North River Steamboat. He conducted the first trials of the completed boat on August 9, 1807 (four years to the day since his successful demonstration in Paris), and it began its first, historic trip up the Hudson on August 17. The design was efficient, but still in need of some refinement; it had only two covered cabins for passengers, at the front and rear; the paddle wheels were not enclosed, thus producing a lot of spray behind them; the steam engine was in the middle of the boat, uncovered and in full view; and the tiller used to steer was at the back of the vessel—the crewman in charge of navigation had difficulty seeing where the boat was going.

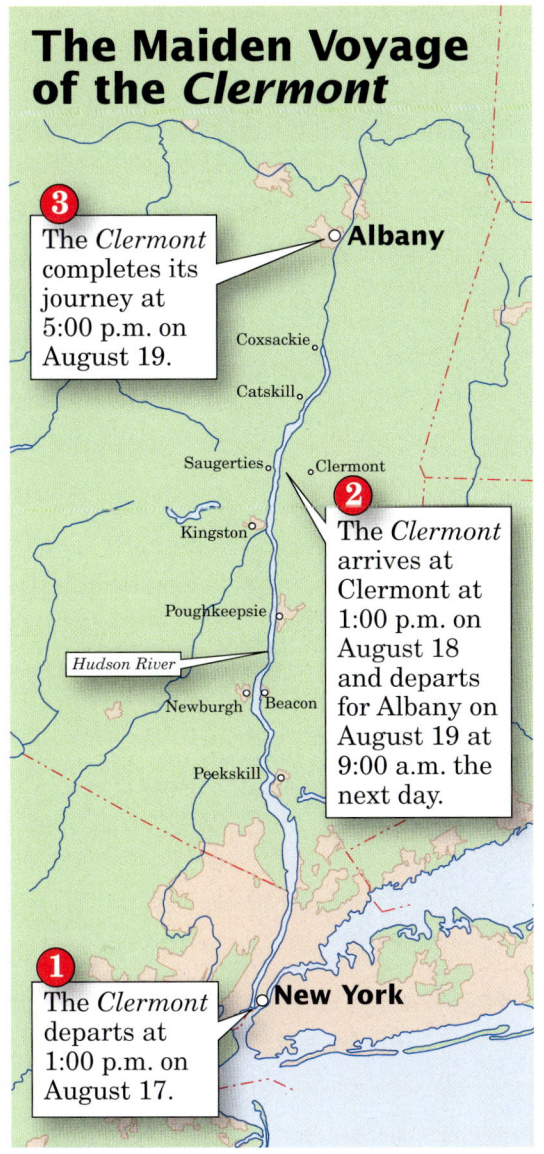

The Maiden Voyage of the *Clermont*

3 The *Clermont* completes its journey at 5:00 p.m. on August 19.

2 The *Clermont* arrives at Clermont at 1:00 p.m. on August 18 and departs for Albany on August 19 at 9:00 a.m. the next day.

1 The *Clermont* departs at 1:00 p.m. on August 17.

The Clermont *made a successful trip up the Hudson River in 1807.*

However, to the disbelief of its passengers, who all seemed to be expecting a failure, the boat plowed on hour after hour, making steady progress upstream toward Albany. After a stop at Clermont, New York, on the second night, the boat arrived at Albany on August 19, having completed the 150-mile (240-km) journey in 32 hours at an average speed of about 4.7 miles per hour (7.5 km/h). The following day it began the return journey to New York City, completing it in 30 hours.

Fulton had proved that running a regularly scheduled—and therefore commercially viable—steamboat passenger service was possible. It was the greatest triumph of his life. He soon set to work improving the boat's steering arrangements and passenger accommodation (including sleeping quarters), and the service became a popular and profitable venture. Fulton and Livingston soon built two other boats, the *Car of Neptune* and the *Paragon*, to join the *Clermont* on the same route. Even before this success, however, Fulton had begun thinking about how steamboats could be used to open up the new territory that the United States had acquired in the Louisiana Purchase of 1803. Through Livingston's influence in government, the two men were able to obtain a monopoly on steamboat operations both in the state of New York and on the lower end of the Mississippi River.

To run on the Mississippi, Fulton and Livingston's partnership had a steamboat, the *New Orleans*, built in Pittsburgh with a 30-horsepower Boulton and Watt engine. Launched on the Ohio River late in 1811, the ship had difficulty navigating downriver from Louisville, Kentucky, continually running into sandbars on the broad, shallow Ohio and Mississippi rivers. It eventually arrived in New Orleans, Louisiana, three months after its launch, but from then on it was restricted to running between New Orleans and Natchez, Mississippi. It ran aground and sank in 1814. Clearly boats used to navigate midwestern rivers needed to be designed differently (broader across the base and able to navigate even shallower waters) from those used on midatlantic rivers. Fulton left this challenge to others.

TIME LINE

1765	1782	1786	1788	1794
Robert Fulton born in Lancaster, Pennsylvania.	Fulton moves to Philadelphia to work as a painter.	Fulton travels to London to study painting.	Fulton begins experiments with steamboat design.	Fulton receives a British patent for a canal system.

Fulton the Artist

Although he is known as an inventor and engineer, for the first half of his life Robert Fulton thought of himself mainly as an artist. He was self-taught in his early years, and his natural ability to draw plans and diagrams of machines also gave him the skill to make portraits, working in a restrained, realist style. He went to London with a letter of introduction from Benjamin Franklin (1706–1790), who had become a friend in Philadelphia, to Benjamin West (1738–1820), the Pennsylvania-born artist who was one of the painters to King George III. West helped Fulton refine and polish his technique.

The largest artwork that Fulton ever painted was in Paris. He had almost certainly seen the great panorama building opened in Leicester Square, London, in 1793 by Robert Barker. Barker was the inventor of the panorama, a circular, 360-degree painting of a landscape or cityscape, viewed from a central platform to give the same perspective a person would have on the spot.

Shortly after Fulton moved to Paris in 1797, he persuaded a fellow American who had recently bought property in the city to build his own circular panorama building on it (the street on which it was located is still called the *Rue des Panoramas*). Fulton's panorama, which opened in 1800, showed a view of Moscow being destroyed by fire (*L'incendie de Moscow*)—one of many fires in which the wooden city had burned down. Fulton used the profits from the entrance fees to the panorama to help fund the building of his submarine, the *Nautilus*.

TIME LINE

1797	1802	1803	1807	1815
Fulton travels to Paris to build weaponry for the French government.	Fulton enters partnership with Robert R. Livingston to build steamboats.	Fulton successfully demonstrates his steamboat.	Fulton conducts the first trial of the *Clermont*.	Fulton dies.

LATER YEARS

In his last two major steam-vessel projects, Fulton returned to an idea that the early Scots steam enthusiast Patrick Miller (1731–1815) had advocated: a paddle-wheeled catamaran—a ship made from two hulls lashed together side by side, with the paddle wheel in the space between them rather than on either side. When war was declared with Britain (the War of 1812), Fulton had the idea of applying this design to a warship, where its great advantage would be that the paddle wheel's position between the hulls would protect it from the fire of enemy guns.

The ship in which he put this idea into practice was called the *Demologos*. Built to protect New York harbor, it was essentially a floating platform for 30 large cannons, plus two cannons of Fulton's own design that could fire a 100-pound (45-kg) shot underwater over a short range. Like his ferry design, it had rudders at either end of both hulls, and its paddle wheels could turn in reverse, allowing it to move both forward and

Modern steamboats on the Mississippi River near New Orleans, Louisiana.

back with equal ease. Fulton worked on this design through 1813, but construction was not actually authorized until the middle of 1814. Although building then proceeded very quickly, the ship was not finished by the time the peace treaty ending the war was signed on December 24, 1814, and it never entered into active service. Several decades would pass before steam power was widely adopted by the world's navies.

In February 1815 Fulton got soaking wet on a journey to inspect the *Demologos*'s construction in New Jersey; he ignored the cold that resulted and carried on working, but shortly afterward developed pneumonia, of which he died on February 23, 1815. Both houses of Congress wore mourning clothes to honor the great engineer.

Two centuries later, it is clear how far ahead of his time many of Fulton's ideas were, particularly for submarines and steam-powered warships. In his own day, he was most honored for the breakthrough that gave a huge boost to transportation and trade all over the world: the establishment of steamboats as a fast new means of moving people and freight along rivers and coasts.

—Jonathan Dore

Further Reading

Books
Sale, Kirkpatrick. *The Fire of His Genius: Robert Fulton and the American Dream*. New York: Free Press, 2001.

Sutcliffe, Andrea. *Steam: The Untold Story of America's First Great Invention*. New York: Palgrave Macmillan, 2004.

Web sites
Robert Fulton and the *Clermont*
  Full text of a 1909 biography of Fulton by Alice Crary Sutcliffe.
  http://www.ulster.net/~hrmm/diglib/sutcliffe/preface.html

Steamboats.com
  A page providing links to photos, articles, and research about steamboats.
  http://www.steamboats.com

Steam Engine Library
  A collection of historical documents relating to the history of the steam engine.
  http://www.history.rochester.edu/steam

See also: Bushnell, David; Ericsson, John; Military and Weaponry; Transportation.

ASHOK GADGIL

Inventor of ultraviolet water
purification system
1950–

Ashok Gadgil is one of the leading scientists in the United States, having made his mark by inventing low-cost and efficient solutions to some of the world's most devastating problems, most notably unclean drinking water. One of his inventions, the UV Waterworks system, is used daily to purify drinking water for hundreds of thousands of people in Mexico, the Philippines, and other areas where the water is dangerous, even deadly, to drink. His inventions reveal a dedication to developing technology that improves the quality of life, especially for the poor, around the globe.

EARLY YEARS

Ashok Gadgil was born in Bombay, India, in 1950. Like many other children who grew up to be great scientists, he displayed curiosity from an early age about the way things worked. By the time he was in the fourth grade, he had read all of the science textbooks at his local high school. To supplement his knowledge, he read magazines like *Popular Science*, *Popular Mechanics*, and *Scientific American*.

When it was time to pick an area of study, Gadgil chose physics, instead of his parents' preference, medicine. He earned a bachelor's degree in physics at the University of Bombay in 1971. Disenchanted with the theoretical nature of his first degree, he pursued a master's degree in applied physics at the Indian Institute of Technology at Kanpur, from which he graduated in 1973. That year, he traveled to the United States to study at the University of California–Berkeley, where he earned a PhD in 1979 and became an expert on solar energy and heat transfer. In 1980, he began working at the Lawrence Berkeley National Laboratory (LBL), the oldest of the U.S. Department of Energy's national laboratories, as a staff scientist in the energy and environmental technology division.

AN EYE TOWARD HOME

After Gadgil spent three years at LBL, he and his wife moved back to India. Hoping to bring some of the latest ideas in energy, technology, and conservation to his homeland, Gadgil began work at a research

Ashok Gadgil poses with his invention, the UV Waterworks system (undated photograph).

A UV Waterworks system installed in Bhupalpur, India. The hand pump feeds the blue plastic tank; the cement structure is opened for routine maintenance of the UV unit.

institute in New Delhi. He tried to find solutions for basic problems, including the most efficient way to heat water and homes. In five years, he was granted four patents for solar heaters, and he made strides in electrical power conservation. However, growing frustrated by the Indian government's placing bureaucratic impediments in his path each time he tried to implement a change, he returned to LBL in 1988.

Gadgil's attention was again drawn to India in 1993, when a deadly outbreak of cholera spread throughout the southeastern part of the country. He was reminded of his childhood, when several cousins died from similar waterborne diseases. The quest for safe, clean water became the environmental issue at the forefront of Gadgil's work.

THE WATERWORKS

The cholera epidemic, which spread from India to Bangladesh and Thailand, killed up to ten thousand people in a matter of months. In addition to the lives lost, the epidemic highlighted a problem that already existed around the world. Waterborne diseases, including cholera, are one of the leading causes of death in many parts of the developing world, particularly sub-Saharan Africa and parts of Asia. An estimated two million people—the vast majority of them children—die from such diseases each year.

Gadgil was deeply affected by the deaths in his homeland. He wanted to develop a cheap, safe, and sturdy water purification system that could operate with little or no energy. The two most common ways of purifying water in the developing world had significant limitations: one method, using chlorine, required training as well as access to chlorine bleach. The other, boiling water, required a great deal of heat, using vital natural resources and causing pollution and deforestation. A third option, using ultraviolet (UV) light, had existed since the end of the 19th century, but no practical way for it to be used in poor or rural areas had been developed. Most UV water purification systems required pressurized water, which worked well in areas with reliable supplies of electricity. Poorer areas, however, required a system that could work with little or no electricity.

Gadgil's invention, now known as the UV Waterworks system, is approximately the size of a microwave oven. It could clean water at a rate of four gallons (18 l) per minute and could provide clean, safe drinking

In 2005, children in the village of Kothapeta, India, fill bottles with purified water from a UV Waterworks tap.

TIME LINE

1950	1979	1980	1983	1988	1993	1996
Ashok Gadgil born in Bombay, India.	Gadgil earns a PhD from the University of California-Berkeley.	Gadgil begins work at a U.S. Department of Energy laboratory (LBL).	Gadgil returns to India to work at a research institute.	Gadgil returns to LBL.	Gadgil begins to develop the UV Waterworks system.	Gadgil receives several awards for his invention.

water for only pennies per person per year. One unit could serve up to one thousand people for roughly $70 per year. Aside from bringing safe water to countless people around the world, it had an added environmental benefit. If used instead of boiling water, which required burning two to three tons (1.8 to 2.7 metric tons) of firewood daily, the UV Waterworks system could prevent the production of up to a ton (.9 metric ton) of carbon dioxide in the air each day.

The importance of Gadgil's invention was quickly recognized. In 1996, he received the Discover Award for the most significant environmental invention of the year, and the Popular Science award for "Best of What's New—1996." As use of the UV Waterworks expanded, reports of its success started to pour in. In one area in Mexico, for example, the incidence of diarrhea dropped by 93 percent.

OTHER INVENTIONS

Water purification is just one of Gadgil's pursuits as an inventor. He has developed various other safety and environmental devices, including an airvest that keeps toxic fumes away from painters, a space heater suitable for people living in the Himalayas, a water heater for use in cities in India, and a low-environmental-impact energy plant in Brazil. He has been called on to devise ways to protect U.S. citizens in the event of a chemical or biological terrorist attack. He has also helped refugee women in Darfur, Sudan, develop more efficient cooking tools to reduce time spent gathering wood—a time when a woman's risk of assault and rape in war-torn areas is extremely high. Gadgil has also worked to develop a cheap and effective way to remove arsenic, a deadly poison, from the drinking water in Bangladesh.

Gadgil has been extensively recognized for his tireless work in improving the living conditions of people around the world. In 1991, he was awarded the Pew Fellowship in Conversation and the Environment for his work increasing energy efficiency in developing countries. In 2002, he won the World Technology Award for Energy, and two years later he won the Tech Award Laureate in Health from the Tech Museum of Innovation, in San Jose, California. Gadgil continues to lead a team of scientists at LBL who are particularly concerned with airflow and pollutants. He was one of seven scientists featured in *Me and Isaac Newton*, a full-length documentary about motivation and inspiration, released in 1999 by the noted director Michael Apted.

—Laura Lambert

Further Reading

Book

Brown, David E. *Inventing Modern America: From the Microwave to the Mouse.* Cambridge, MA: MIT Press, 2001. Excerpt available online at:
http://web.mit.edu/invent/www/ima/gadgil_intro.html

Web sites

Innovative Lives: "UV Waterworks: Ashok Gadgil"
A profile of Ashok Gadgil from the Lemelson Center for the Study of Invention and Innovation.
http://invention.smithsonian.org/centerpieces/ilives/uvwater.html

UV Waterworks
The home page of the UV Waterworks system.
http://eetd.lbl.gov/iep/archive/uv/

See also: Environment and Inventing.

GALILEO GALILEI

Developer of the telescope

1564–1642

Many inventions have changed history, but few have had such a dramatic impact as the telescope. When telescopes were first used to look at the night sky in the early seventeenth century, people were forced to question how the universe worked. Such exploration upset fundamental religious ideas and marked the beginning of the scientific age. The person responsible for this upheaval was Galileo, an Italian physicist and mathematician who is often considered to be the father of modern science.

EARLY YEARS

Galileo Galilei (usually referred to simply as Galileo) was born in the Italian town of Pisa on February 15, 1564, to Vincenzo Galilei, a musician, and Giulia Ammannati. Vincenzo Galilei had helped to revolutionize musical styles at the end of the Renaissance. He also carried out experiments showing how the length and tension (tightness) of the strings of a musical instrument could produce higher or lower notes. Some historians believe Galileo observed and helped his father, setting the stage for his own scientific experiments.

For the first eight years of his life, Galileo was schooled at home by a private tutor and family friends. In 1574, at age 10, he entered the monastery of Santa Maria di Vallombrosa, perched on a wooded hillside near the city of Florence. Galileo loved the discipline of religious life and planned to be a monk. His father, however, wanted him to be a doctor, so in September 1581, Galileo began attending the University of Pisa. There he studied science and mathematics and began developing his early ideas about gravity. While watching lamps swinging from the ceiling of Pisa's cathedral in 1583, he made one of his most famous observations. Using his pulse as a clock, he found that a pendulum (a swinging mass) always takes the same time to move back as to move forth, no matter how heavy it is. In 1586, he made his first invention: the *bilancetta*, or hydrostatic balance, a device for measuring the density of an object.

FROM PISA TO PADUA

In 1589, Galileo was named a professor of mathematics at Pisa. He soon began to challenge current scientific ideas—ideas that had been established by an ancient Greek scientist and philosopher, Aristotle (384–322 BCE). Galileo proved that objects of different weights, like a cannonball and a feather, fall at the same speed; the feather reaches the ground later only because it is slowed by air resistance. This was the opposite of what Aristotle had taught.

A portrait of Galileo from around 1610 (artist unknown).

An artist's rendering of Galileo's gravity experiment at Pisa, in which he proved that different objects fall at the same speed.

Even at this early stage of his career, Galileo knew he was upsetting people. He chose not to publish some of his arguments against Aristotle, knowing that they were unfinished and extremely controversial. Still, his challenging approach made him enemies. Partly for this reason, he decided to leave Pisa. In 1592, he accepted a new job as professor of mathematics at Padua, where he would remain for the next 18 years. At Padua, he continued to develop his ideas about gravity, working out how missiles flew through the air in an inverted-U-shaped path called a *parabola*. A year after arriving at Padua, he developed another useful invention, a horse-powered pump for raising water, for which he received a patent in 1594.

During a trip to Venice, he met Marina Gamba, who became his housekeeper. Although the couple never married, they had two daughters, born in 1600 and 1601; and a son, born in 1606.

> In questions of science, the authority of a thousand is not worth the humble reasoning of a single individual.
> —Galileo

A STAR IS BORN

On October 10, 1604, a bright new star appeared in the sky over Padua, and it captured Galileo's imagination. He made some calculations and discovered that the star was far beyond the earth's moon, in the part of the sky then known only as "the heavens." This, he realized, was another example of Aristotle's errors. Aristotle had argued that the heavens were perfect and unchanging, but the appearance of a new star meant they could alter.

Although a mathematician by training, Galileo was becoming more interested in astronomy. A few years earlier, he had studied the way ocean tides change at different times of the year. He felt part of the explanation was that earth must be moving. Aristotle, however, had argued that earth was the still center of the universe. Several hundred years after Aristotle, an astronomer known as Ptolemy had based the science of astronomy on

The Sun and the Earth

It was natural for ancient astronomers to assume earth was the center of the universe; as they stared into the sky, they had no sense that earth was moving. The sun, on the other hand, seemed to glide across the sky in an arc. Aristotle (384–322 BCE), the greatest scientist of ancient Greek times, was certain that earth was the still center of the universe. Around 340 BCE, he wrote a book, *On the Heavens*, in which he described the sun, moon, stars, and planets as all revolving around earth in circular paths.

Several hundred years later, an Egyptian astronomer and mathematician, Claudius Ptolemaeus (100–170 CE), or Ptolemy, greatly extended Aristotle's ideas. When he was about 50, he published an enormous, 13-volume astronomical encyclopedia, the *Almagest* ("the greatest"). Like Aristotle, Ptolemy maintained that earth was the hub of the universe and that everything else cycled around it.

One of the first to challenge this view was Polish astronomer Nicolaus Copernicus (1473–1543). Copernicus found many of Ptolemy's assumptions about the universe clumsy. For a better explanation of how the universe worked, Copernicus went back to the ideas of a Greek thinker, Aristarchus (ca. 310–250 BCE), who had suggested the sun was the center of the universe. Copernicus extended this idea into a complete theory of the universe—*De revolutionibus orbium coelestium* (*On the Revolutions of the Heavenly Spheres*). The book was so controversial that it was not published until Copernicus lay on his deathbed in 1543. When Galileo looked through his telescope, he saw not just the moons, the stars, and the planets, but the truth and wisdom of what Copernicus had proposed.

this assumption. Many religious precepts and doctrines had also been based on the idea—known as the geocentric (earth-centered), or Ptolemaic, theory. Now Galileo found himself supporting the very different idea that had been put forward by a Polish astronomer, Nicolaus Copernicus (1473–1543): the sun was the center of the universe and earth revolved around it. This was called the heliocentric (sun-centered), or Copernican, theory (see box, The Sun and the Earth).

Although Galileo believed the earth moved around the sun, he was not confident enough to promote the idea in public. The risks of doing so were enormous. A few years earlier, the fiery Italian philosopher Giordano Bruno (ca. 1548–1600) had boldly championed the heliocentric idea. He was imprisoned for challenging the Catholic church and, after refusing to recant, was burned at the stake.

A NEW WAY OF SEEING

During a visit to Venice in the summer of 1609, Galileo heard about the invention of the telescope, a viewing tube developed by Dutchman Hans Lippershey (ca. 1570–1619) that could make distant objects appear

A composite photograph of the four moons of Jupiter discovered by Galileo (from top left: Europa, Callisto, Ganymede, and Io), taken by the Voyager 1 *spacecraft in 1979.*

How Telescopes Work

"What shall we make of this? Shall we laugh, or shall we cry?"

Anyone who has ever used a magnifying glass knows that a lens (a single, curved piece of glass or plastic) can make objects in the world look bigger and closer. Putting two lenses side by side makes a pair of eyeglasses, but putting one lens in front of another has a different effect: it makes objects that are far away seem closer. Galileo's telescope was not much more sophisticated than this. It had a pair of lenses fixed at opposite ends of a long tube. The lens at the distant end gathered light rays from far away and brought them to a focus inside the tube; the lens at the near end spread the rays out again to make a bigger and clearer image.

Using a combination of mathematical analysis and careful experimenting, Galileo quickly figured out the best way to make a telescope. His first instrument could magnify things only three times, but he soon built better telescopes that could make things look eight and then 20 times bigger. These powerful instruments enabled Galileo to make remarkable discoveries and provided some of the first evidence that earth rotates around the sun. One reason Galileo's ideas were initially so surprising and controversial was that no one else had made such powerful telescopes, so others could not immediately verify his claims.

Yet Galileo's ideas proved controversial even after others had confirmed them. Since the telescope had been the source of these ideas, many people distrusted it. Giovanni Magini, professor of mathematics at Bologna, found himself unable to see anything through the telescope, even when Galileo helped him. Cesare Cremoni and Giulio Libri, two professors at Padua and Pisa, simply refused to look through a telescope at all, prompting Galileo to wonder, "What would you say of the learned here, who . . . have steadfastly refused to cast a glance through the telescope? What shall we make of this? Shall we laugh, or shall we cry?"

Galileo's English counterpart, Isaac Newton, extended many of Galileo's theories on motion, gravity,

Replicas of telescopes invented by Galileo (right) and Newton (left).

and astronomy. Thirty years after Galileo's death, in 1672, Newton invented a better telescope. Unlike Galileo's telescope, which is called a refractor because it uses lenses to bend, or refract, light, Newton's telescope used mirrors. Telescopes like Newton's are known as reflectors, because they reflect light instead of refracting it. Reflecting telescopes can be made more powerful than refractors because large mirrors are easier to make than large lenses. Although telescopes have changed greatly since Galileo and Newton first developed them, most still work using reflection or refraction.

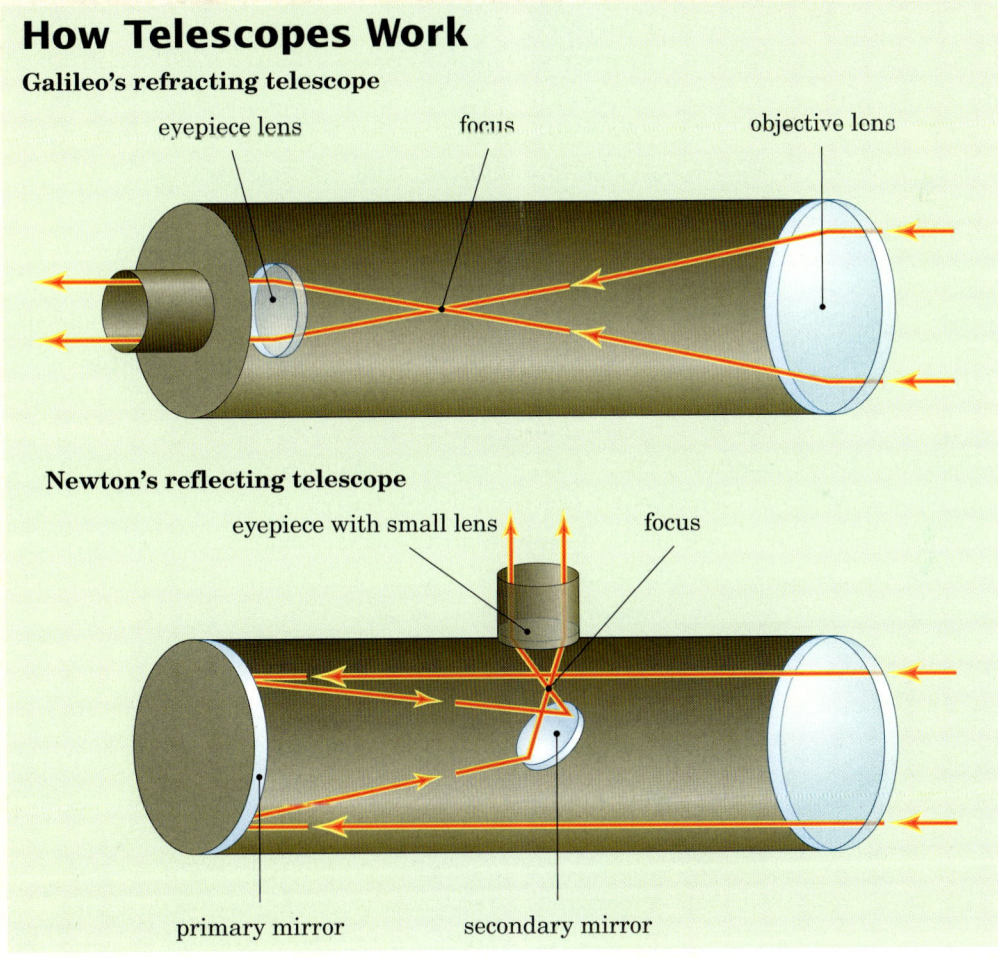

Galileo's original instrument used lenses to bend or refract light and is therefore called a *refracting telescope*. The simplest refracting telescope (or refractor) has two lenses. The main or objective lens gathers light from distant stars and brings an image into sharp focus. The smaller eyepiece lens magnifies the image. Newton built the first successful reflecting telescope using mirrors instead of lenses. In place of the objective lens, there is a large primary mirror. This mirror gathers light from distant stars and brings it to a focus. A smaller, secondary mirror reflects this light toward the viewer, and a small lens magnifies the image.

closer. Although the invention was secret, Galileo figured out the science for himself and made his own powerful telescopes. Many people correctly thought telescopes would revolutionize navigation at sea, but for Galileo the prize was bigger: he could now look directly into the heavens to understand the universe.

Over the next few months, Galileo used his telescope to make a series of remarkable discoveries. In November and December of 1609, he saw that the earth's moon was mountainous and not smooth, as others had believed. In 1610, he discovered four moons of Jupiter, observed Saturn and Venus, and saw the Milky Way, the galaxy to which earth belongs. These findings caused a sensation in Italy and secured Galileo's fame as a great scientist. When he published his results in a book titled *The Starry Messenger* in March 1610, he dedicated it to Cosimo II, the grand duke of Tuscany. The duke returned the honor four months later by giving Galileo a prestigious new job as his personal philosopher and mathematician. With this appointment, Galileo moved from Padua to Florence, where he had much more time for research and study.

CHALLENGING THE CHURCH

Galileo realized that what he could see with his telescope could not be explained by the old geocentric theory. If he assumed that the earth was moving around the sun, however, as the heliocentric theory suggested, his observations started to make sense. Leaders of the Catholic church now began to take more interest in Galileo's work. Although they confirmed the discoveries he had made, they did not agree with his conclusion that the earth must be moving. As Galileo promoted this idea, he made more and more enemies.

> Two of Galileo's lesser-known inventions were a machine for picking tomatoes and a hair comb that doubled as a knife and fork.

Galileo clearly saw that a major crisis was coming. Beginning in 1613, he wrote a series of long letters defending his ideas. One of these, to the dowager grand duchess Christina of Lorraine, written in 1615, was widely circulated. In it, Galileo defended himself against enemies whom he claimed were determined to destroy him because "I hold the sun to be situated motionless in the center of the revolution of the celestial orbs while the earth revolves around the sun." Yet he also claimed that this was not an argument against the Bible, which suggested that the earth was the center of creation. Galileo argued that the Bible was more of a general guide than something that should be read as the literal truth. It was also not a textbook on astronomy: "The intention of the Holy Ghost is to teach us how one goes to heaven, not how heaven goes."

Early in 1616, church leaders summoned Galileo to Rome to be questioned on his views. Shortly afterward, at the personal request of

Engraved plate of astronomers at work, from the book Rosa Ursina *(1630), by German astronomer Christoph Scheiner.*

Pope Paul V, the church formally warned him: he must stop promoting his theory that the earth moved as the truth, although he was still free to discuss it as an idea. Galileo appeared to heed the warning. For seven years, he worked quietly in Florence, considering a range of scientific and mathematical problems. After studying comets, he published a new book, *The Assayer*, in 1623, dedicating it to the new pope, Urban VIII. The pope was enthusiastic, but an unknown person complained that *The Assayer* was another challenge to the church. After an investigation, Galileo was cleared.

TRIAL AND IMPRISONMENT

Before he became Pope Urban VIII, Cardinal Maffeo Barberini had been a friend and supporter of Galileo. This connection seemingly gave Galileo more confidence to promote his ideas. In 1624, he started work on a new book, having gained permission from the pope, with the stipulation that Galileo reach no conclusions at odds with the church's teachings. The pope even gave Galileo a pension in 1630. However, in 1632, when Galileo published his book, *Dialogue Concerning the Two Chief World Systems, Ptolemaic and Copernican*, he left little doubt about his views.

The pope was in a difficult position. Just before the book was printed, he asked church leaders to examine what Galileo had written. They summoned Galileo to Rome and put him on trial in April 1633, charging him with disobeying the pope's orders not to promote the Copernican theory. Galileo admitted that he had been wrong to support the Copernican theory too strongly and offered to write a new book reaching a different conclusion. However, the pope and other church leaders decided to imprison him; in June 1633, Galileo was sentenced to house arrest for the remainder of his life.

An undated illustration of Galileo facing accusations of heresy.

Galileo lived another eight years. Despite confinement and failing health, he wrote another important book, *Dialogues Concerning Two New Sciences*. This discussion of how objects move led English scientist Isaac Newton (1643–1727) to form important ideas about motion and gravity some years later. In 1637, Galileo made further discoveries about the earth's moon with his telescope—but he lost his sight soon afterward. His mind, however, remained brilliantly sharp. In his last year of life, now completely blind, he started thinking again about his first scientific discovery: the way pendulums swing. After a little experimentation, he invented a way of using a pendulum to keep time in a clock using a

TIME LINE

1564	1589	1592	1604	1610
Galileo Galilei born in Pisa, Italy.	Galileo takes position as professor of mathematics at University of Pisa.	Galileo takes position as professor of mathematics at University of Padua.	Galileo sees a new star and begins serious study of astronomy.	Galileo discovers four moons of Jupiter; publishes *The Starry Messenger*.

NASA artist's rendering of the Galileo *probe passing by Jupiter and one of its moons in 1989.*

gear mechanism called an escapement. He died several months later at his home in Arcetri, Florence, on January 8, 1642.

Combining cleverly designed experiments and careful observations with mathematics and analysis, Galileo was a very modern scientist. He rejected outworn theories in favor of what he could demonstrate in his own laboratory and see with his own eyes. The telescope helped Galileo see a new world—a world of scientific promise that revolved around the sun. The Catholic church appeared to accept its mistake 350 years later, in October 1992, when Galileo was officially pardoned by Pope John Paul II.

—Chris Woodford

TIME LINE

1616	1623	1632	1633	1642
Galileo warned by church leaders to stop promoting heliocentric universe theory.	Galileo publishes *The Assayer*.	Galileo publishes *Dialogue Concerning the Two Chief World Systems, Ptolemaic and Copernican*.	Galileo sentenced to house arrest by church officials.	Galileo dies.

Further Reading

Books

Hightower, Paul. *Galileo: Astronomer and Physicist*. Hillside, NJ: Enslow, 1997.

MacLachlan, James. *Galileo Galilei: First Physicist*. New York: Oxford University Press, 1999.

Panchyk, Richard. *Galileo for Kids: His Life and Ideas*. Chicago: Chicago Review Press, 2005.

Woodford, Chris. *Gravity: Routes of Science*. Farmington Hills, MI: Blackbirch, 2004.

Web sites

Galilean Library
    A detailed archive of Galileo's life.
    http://www.galilean-library.org/

Galileo's Battle for the Heavens
    Information, activities, and resources from PBS.
    http://www.pbs.org/wgbh/nova/galileo/

Galileo Project
    Biography, time lines, and lesson plans about Galileo's life and work.
    http://galileo.rice.edu/

See also: Hooke, Robert; Lippershey, Hans; Newton, Isaac; Optics and Vision; Science, Technology, and Mathematics.

IVAN I. GETTING

Inventor of the Global Positioning System
1912–2003

Ivan Getting envisioned what would become an entirely new form of navigation, the Global Positioning System (GPS), originally designed for military applications. The accuracy and ease of use of GPS have made it popular among civilians as well.

EARLY YEARS

Ivan Getting was born in New York City in 1912. His family was originally from what is now Slovakia. Getting grew up mainly in Pittsburgh, Pennsylvania, but he spent the year 1919 living and attending school in Bratislava, in what was then Czechoslovakia.

Getting won a scholarship to the Massachusetts Institute of Technology (MIT) in Cambridge. Graduating in 1933 in the midst of the Great Depression, he was unable to find a job, so he applied for a Rhodes scholarship to Oxford University in England. He was awarded a scholarship and earned his doctorate in astrophysics from Oxford in 1935.

Getting decided to return to the United States, accepting an offer to become a junior fellow at Harvard University in Cambridge, Massachusetts. While working there on a problem involving cosmic rays, Getting developed a type of circuit called a flip-flop, so named because it can go back and forth between two states, to help him with his work. His circuit was much faster than any other flip-flop then in use, and as computers developed, Getting's circuit would end up being widely employed to create what is known as temporary memory (the impermanent memory that computers use to function when they are on).

THE RADLAB

Getting, however, was moving on to yet another field: nuclear physics. His reasons for choosing this field were rooted in the real world: in 1939, physicists in Nazi Germany split the atom, paving the way for the development of the atomic bomb. By 1940, Nazi Germany had conquered almost all of western Europe.

Although the United States would not formally enter World War II until December 1941, by 1940 the U.S. government, convinced that conflict with Germany was inevitable, undertook a major push into weapons research in con-

An undated photograph of Ivan Getting in his laboratory at the Massachusetts Institute of Technology.

A technician tracks aircraft by radar in 1944.

junction with various universities. Getting was approached in the fall of 1940 by an acquaintance who asked if he wanted to join a new laboratory at MIT that was going to create new forms of defense technology. Getting jumped at the chance—even though he was unsure exactly what work at the lab would involve.

Getting joined the Radiation Laboratory, or RadLab, at MIT. The RadLab focused on developing and improving radar, which was poorly understood at the time. Radar (radio detection and ranging) uses radio waves to calculate the position of objects. Getting directed the laboratory's army radar division and contributed to many of the RadLab's projects, most notably the development of a radar system to target the German V-1 and V-2 flying bombs.

The V-1 (*Vergeltungswaffe* 1, or "retaliation weapon 1") and V-2 were guided, unmanned bombs. As the bombs were small and very fast, conventional antiaircraft fire was almost useless against them. Beginning in mid-1944, Nazi Germany hammered Great Britain, in particular London, with V-1 and V-2 attacks. Getting helped develop a radar tracking system that would detect and automatically follow a target. Such aiming systems proved devastatingly effective against V-1 and V-2 bombs.

AFTER THE WAR

In 1945, Nazi Germany was defeated and World War II came to an end. Getting remained at MIT as a professor of electrical engineering until

1950, when the outbreak of the Korean War made him decide to focus again on defense work.

Getting worked for the U.S. Air Force until 1951, when he joined the defense contractor Raytheon as vice president for research and engineering, a position he held until 1960. At Raytheon, Getting helped develop the Sparrow III and Hawk radar-guided missile systems, which are still in use.

THE SPACE PROGRAM

In 1960, Getting resigned from Raytheon to become the founding president of the Aerospace Corporation, a nonprofit research organization in El Segundo, California, established to help the U.S. Air Force with its space programs. Aerospace made many contributions to the space program, including helping to develop the Mercury and Gemini missions.

In the late 1960s, Aerospace was slated to help create a manned space station for the air force. Getting geared his company up for the effort, hiring hundreds of new people. Then, in 1969, the station was suddenly canceled, leaving Aerospace's new staff with nothing to do.

A committee had been established to decide on the direction of the U.S. space program. Getting suggested several projects to the committee in hopes that the projects would lead to new work for Aerospace, thus preventing layoffs. Among his suggestions was a new system of navigation, the Global Positioning System (GPS).

GPS

Getting would later say that the idea of GPS emerged from his experiences during World War II, which made him realize the importance of knowing exact locations in combat. Ground-based navigation systems can be blocked by natural formations such as mountains and hampered by bad weather; in addition, the curvature of the earth limited their range.

Space satellites opened up the possiblity of a new global navigation system. The U.S. Department of Defense had decided that all the military branches should be on a single navigational system, and a global, weatherproof system like GPS seemed ideal.

Day-to-day development and management of GPS during the 1970s was the responsibility of Colonel Bradford Parkinson of the air force. Getting remained involved nonetheless. By this point in his career, Getting was enormously respected within the military community for his many contributions to defense technology, and he used his stature to ensure that the very expensive GPS project was kept on track and not allowed to languish.

How GPS Works

"Knowing exactly where you are is extremely important."

Getting developed GPS because his experiences in World War II had taught him the importance of location. "Position in the military sense is very essential," he said. "Knowing exactly where you are is extremely important. Knowing exactly where the enemy is [is] important."

GPS has two major components: satellites in space and receivers on earth. Communication between the satellites and the receiver informs a user of his or her position. Each GPS satellite is locked into a precise orbit around the earth so that at any given time four satellites are in reach of a receiver from any point on earth. The satellites emit radio signals to the receivers, which are programmed with a guide telling them where each of the satellites is in space.

To determine a user's location, the receiver calculates the time taken for a signal to arrive from the four satellites. The signals are sent simultaneously, but a signal emitted by a satellite located farther away from a receiver will take longer to reach it. Such brief lags in the signal are enough for the receiver to calculate its distance from each of the four satellites. Once the receiver knows the satellites' locations in space, and its distance from the satellites, it is able to determine its own position on earth.

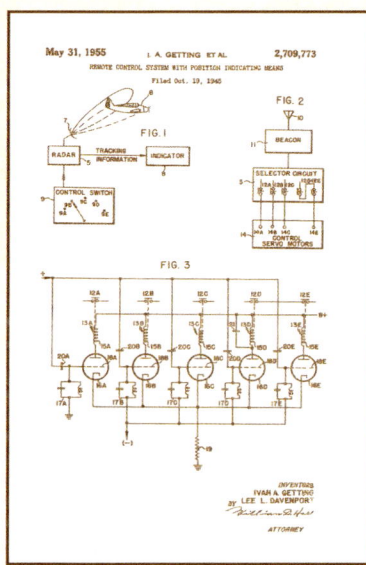

Drawings from Getting's patent application for GPS.

A GPS receiver can pinpoint its exact location on earth by calculating data based on the position of satellites orbiting earth.

GPS was developed for military use and remains an important tool. In 2006, United Nations soldiers landing on a beach in southern Lebanon check their GPS receiver to confirm their location.

In 1978, a year after Getting retired from Aerospace, the first experimental GPS satellite was launched; nine more satellites were launched during the 1980s. In 1989, the first satellite for the current GPS was launched, and by 1994 the last of the 24 satellites was launched, creating what is the Navstar satellite system.

FROM SOLDIERS TO CIVILIANS

Although GPS was a U.S. Department of Defense project, from the beginning it was clear that the system could be very useful for civilians in need of navigational guidance—for example, backpackers or boaters. Making GPS available to civilians could have created a security threat, however, if a hostile army outfitted its troops with GPS receivers. The Navstar system was designed to emit two signals: one to be received by ordinary GPS receivers and one to be received only by the U.S. military's receivers. The military signal was made more accurate, to allow a military receiver to indicate a user's position within 30 to 60 feet (9 to 18 m).

The civilian signal was made less accurate and reported a user's position within 90 feet (27 m). However, military planners worried that the difference in accuracy between the two systems was not great enough to give military signal users the advantage. When the GPS system first became operational in 1993, the civilian signal was degraded further to restrict a civilian GPS receiver to reporting positions within 300 feet (91 m).

As the 1990s progressed, pilots of boats and airplanes began to rely more and more on GPS receivers. The U.S. government became concerned that the degraded civilian signal could lead to serious boat and air-

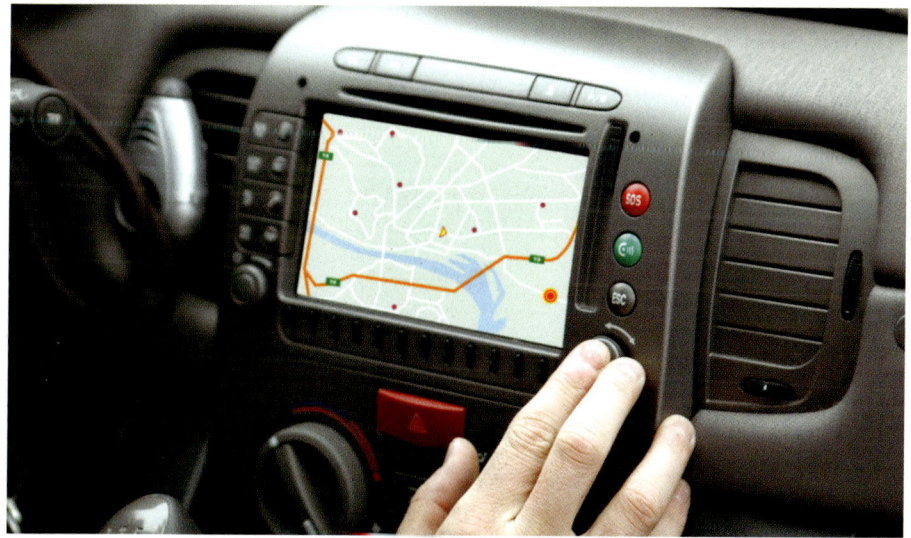

GPS receivers are becoming increasingly common on the dashboards of modern cars.

plane accidents. As GPS receivers became increasingly popular, these worries began to outweigh military concerns. In 2000, the U.S. government restored the accuracy of the civilian GPS signal. If necessary, the signal can again be degraded.

GPS EVERYWHERE

The increased accuracy of civilian GPS, combined with smaller, cheaper receivers, increased the system's popularity with the public. Among the many civilian GPS users was Getting, who in February 2003 shared the Draper Prize, a prestigious engineering award, with Parkinson for their joint work on GPS. Eight months later, Getting died at his home in Coronado, California, at the age of 91.

In the 21st century, civilian GPS receivers are a multibillion-dollar business. By 2004, nearly 30 percent of new cars in the United States were purchased with dashboard GPS receivers already installed, and countless

TIME LINE

1912	1935	1940	1945	1960
Ivan Getting born in New York City.	Getting earns a doctorate in astrophysics from Oxford University.	Getting joins the Radiation Laboratory (RadLab) at MIT.	Getting works as a professor of electrical engineering at MIT.	Getting founds the Aerospace Corporation.

TIME LINE (continued)

1969	1977	1978	1994	2003
Getting begins to develop the Global Positioning System (GPS).	Getting retires from Aerospace.	The first experimental GPS satellite is launched.	The Navstar satellite system is completed.	Getting wins the Draper Prize; he dies later that year.

other drivers have purchased add-on GPS units. Wristwatches and cell phones can be purchased with GPS receivers, and stand-alone receivers can be purchased for as little as $100. Geocaching, a game designed for use with GPS, directs people using their receivers to find hidden prizes. Getting's invention has turned satellite navigation from an esoteric craft to an easy, accessible, and pleasurable technology.

—Mary Sisson

Further Reading

Book
Egbert, Robert I., and Joseph E. King. *The GPS Handbook.* Short Hills, NJ: Burford, 2003.

Web sites
Aerospace Corporation
   The research organization that Getting helped found.
   http://www.aero.org
Research Laboratory of Electronics at MIT
   The modern successor of the RadLab.
   http://www.rle.mit.edu
USNO NAVSTAR Global Positioning System
   An overview of GPS technology by the U.S. Naval Observatory.
   http://tycho.usno.navy.mil/gpsinfo.html

See also: Military and Weaponry.

KING C. GILLETTE

Inventor of the safety razor
1855-1932

When it comes to shaving, 21st-century men have many options: electric razors with pulsating blades, disposable plastic razors purchased by the dozen, and, for the nostalgic, old-fashioned straight razors and shaving brushes. At the end of the 19th century, however, the safety razor was a man's primary choice for shaving at home. Safety razors were safer to use than straight razors, but without constant sharpening on a leather strap, they quickly became dull. Then, in 1901, King C. Gillette, a traveling salesman, developed the disposable safety razor and helped transform not only shaving but also important business practices.

EARLY YEARS

King Camp Gillette was born in Fond du Lac, Wisconsin, in 1855, but moved to Chicago with his family at the age of four. His parents could both be considered inventors of a sort. King's father developed new devices and tinkered with technology as a hobby, and his mother created new recipes in the kitchen. In 1887, she published many of her "inventions" in *The Whitehouse Cookbook*, which remained in print for more than a century.

In 1871, the family lost everything in the great Chicago fire and relocated to New York City. Gillette's father found work as a patent agent and regaled the family with stories of the new inventions that came across his desk. Gillette left school at age 17 to begin a career as a traveling salesman. He also had "a tinkerer's bent," as one journalist put it. By 1890, Gillette had secured patents for a few inventions, but he never had success developing a business with any of them.

At the age of 36, Gillette became a New York and New England sales representative for the Baltimore Seal Company, which manufactured seals for pumping equipment and beverage containers. The president of the company, William Painter, built his fortune on one invention in particular, the Crown Cork, a cork-lined bottle cap. (In fact, Painter later renamed the company the Crown Cork & Seal Company.)

Painter had a profound effect on Gillette's life. He urged Gillette to find his own "Crown Cork"—that is, something that was used once and then thrown away, so that the customer would keep coming back for more. In essence, Painter was pushing Gillette to invent something disposable. Despite some initial skepticism, Gillette took Painter's words to heart, and for years he tried to find the perfect idea.

IN THE BATHROOM

A photograph of King C. Gillette taken by Benjamin Falk around 1904.

In the spring of 1895, Gillette finally found his version of the Crown Cork. As is the case with many other

A 19th-century cartoon illustrates the painful challenges of shaving before Gillette; the caption reads, "Egads, one might as well shave with a SAW!"

great inventions, the idea came to Gillette at an unexpected moment; he thought of it in the bathroom. One morning, Gillette woke and tried to shave, but his Star Safety Razor had become very, very dull. Safety razors of that time had to be sharpened on a leather strap. After a while, however, the blades became too worn to re-sharpen.

Gillette began to imagine a new kind of razor, one that did not need continual sharpening. He later wrote that he wanted something "made cheap enough to do away with honing and stropping and permit the user to replace dull blades by new ones." Gillette envisioned a small, square sheet of steel, fashioned into a thin, double-sided blade. It would be safe, easy to use, cheap, and, on the advice of Painter, disposable.

DEVELOPING THE RAZOR

> The thoughts herein contained are dedicated to all mankind; for to all the hope of escape from an environment of injustice, poverty, and crime, is equally desirable.
>
> —Dedication from Gillette's *The Human Drift* (1894)

Gillette's invention, although simple in theory, proved so complex in reality that six years were needed to develop a workable prototype. Gillette was living in Boston when the idea struck him, so he visited the nearby Massachusetts Institute of Technology (MIT) to discuss his plan with some engineers who worked with steel. Each told him it would be impossible to create a blade as small, hard, thin, and inexpensive as he wanted.

Finally, in 1900, Gillette met engineer and inventor William Nickerson, who had been educated at MIT and who had established a reputation as a skilled inventor. Although Nickerson, like the other scientists, was somewhat skeptical at first, he took on the challenge to make Gillette's blade. They founded the American Safety Razor Company in 1901.

The two men worked in a one-room workshop above a fish market on Boston's waterfront. Gillette and some early investors put about $5,000 into research and development. Nickerson designed a machine that could make inexpensive 40-gauge blades that were 0.88 inch (2.2 cm) wide and just 0.006 inch (0.015 cm) thick. With those specifications, one pound of steel could create almost 400 blades.

Gillette took Nickerson's blades and developed a blade carrier attached to a handle. It was a success. Although Nickerson developed the blade, his name was too close to the word *nick*, which that was so often associated with shaving. So, the Gillette Safety Razor was born; in 1902, Gillette and Nickerson renamed their venture the Gillette Safety Razor Company.

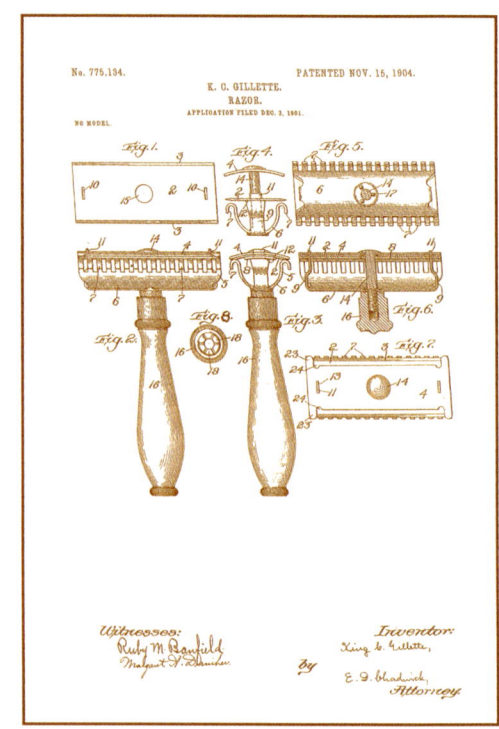

Illustrations from Gillette's 1904 patent for the safety razor.

THE START OF SOMETHING BIG

Production of Gillette razors did not begin until 1903. The company struggled under the weight of more than $12,000 in debt. One acquaintance of the two men told a reporter for the *Wall Street Journal*, "Delays were so serious that Gillette was almost persuaded to shut down and say good-bye to all the rainbows." The sales for that year were meager: 51 razors and 168 blades.

The next year saw a quick reversal. On November 15, 1904, Gillette was granted U.S. Patent No. 775,134, for "certain new and useful improvements in razors." Gillette saturated the market with his product, often selling the razors at a discount or giving them away free. The company sold more than 90,000 razors and 124,000 blades that year. Gillette put his portrait and signature on every package. The product's success ensured that he quickly became one of the most familiar faces of his day. Some have argued that Gillette blades were as recognizable a product as Coca-Cola or Ford automobiles.

In 1905, Gillette expanded abroad, opening a sales office in London. By 1908, the company had production facilities in Canada, England, France, and Germany, in addition to those in the United States. By 1910, Gillette razors dominated the market and Gillette was a millionaire. At a decade old in 1911, the company was selling 35 million blades per year. By 1915, it was selling more than 70 million blades worldwide.

BUSINESS SENSE

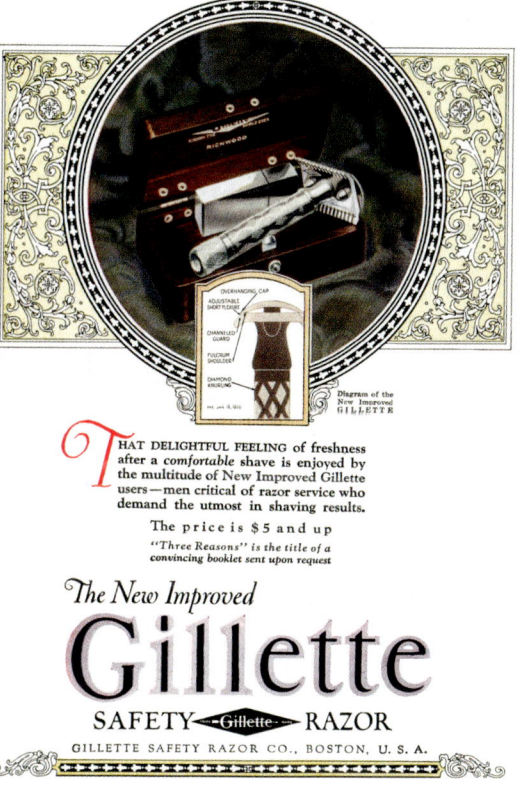

Undated advertisement for an improved razor. Gillette emphasized that advertising was vital to his company's continued success.

The business model that Gillette developed is nearly as famous as the razor itself. Gillette sold many of the razors at a loss, and even gave them away—not just at the beginning of the business but throughout. He realized that giving razors away for nothing or next to nothing would build his market base. With a large base of men using his razor, in the long run, he could more than make up "lost" sales revenue through blade sales. Men might buy one razor in a lifetime, but they would buy dozens of blades each year. The razor is an example of what is now called a loss leader.

Gillette employed this business model again when the United States entered World War I. He developed razor sets for the U.S. Army, which he sold to the government at reduced rates. In return, more than 3.5 million soldiers used his razors and blades and many of them became customers for life.

Advertising also played a big role in the success of the Gillette razor. In a memo from 1912, Gillette told his executives that "the whole success" of the company rode on ads. The company was one of the first to use sports figures to promote razors. Today, sports figures and other celebrities sell not just razors, but a wide range of consumer products, such as shoes, clothing, cars, and kitchen appliances.

ANTICAPITALIST IDEAS

Gillette's incredible success as a businessman is somewhat ironic given his other lifelong passion—the establishment of a socialist utopia. In 1894, before the development of the disposable blade, Gillette wrote a book, *The Human Drift*. An anticapitalist book in which he criticized the business world and called competition the root of all evil, *The Human Drift* also advocated that all industry be taken over by one, supposedly benevolent, corporation. As an antidote to the grimy cities that were born of the Industrial Revolution, Gillette proposed a new kind of world order, a pollution-free, efficient, hive-like commune powered by hydro-energy created by Niagara Falls.

Gillette wrote *World Corporation* in 1910, building on his original ideas. Before World War I, Gillette asked Theodore Roosevelt to serve as leader of a utopian society he hoped to build in what was then the Arizona Territory. Roosevelt declined, as did Henry Ford. Gillette wrote a final book, *The People's Corporation*, in 1924.

LATER YEARS

With the success of the Gillette razor came imitation, and for many years Gillette was embroiled in patent battles. In his later years, he was more a figurehead for the company than its leader. Eventually, he turned his interests to a new project, an unsuccessful attempt to extract oil from shale.

Gillette, who had resigned as president of the company in 1931, died in 1932 in southern California, where he had bought land to grow oranges and dates. He was nearly bankrupt as a result of the Great Depression.

The Gillette Safety Razor Company went on to become remarkably successful. The first electric shaver, the Gillette dry shaver, was introduced in 1938. In time, the company's product line included Foamy

TIME LINE

1855	1872	1895	1901	1903	1931	1932
King Camp Gillette born in Fond du Lac, Wisconsin.	Gillette drops out of school to become a traveling salesman.	Gillette gets idea for a disposable razor.	Gillette and William Nickerson found the American Safety Razor Company.	Production of Gillette razors begins.	Gillette resigns as president of company.	Gillette dies.

The Gillette company continues to emphasize advertisements and celebrity endorsements. Here, Sean "Diddy" Combs introduces a new razor at an event in 2004.

shaving cream, Right Guard antiperspirant, Oral-B dental products, and other, nonpersonal goods, such as pens. Starting in the 1970s, Gillette began introducing new shaving technology. At the turn of the 21st century, the company had annual sales of more than $9.9 billion. Gillette was acquired by Procter & Gamble in 2005.

The roots of such success go back to 1901 and King Gillette. Gillette was a pioneer in disposable consumer products, and he ushered in a new era in commerce with his razor-and-blade business model; it is the idea behind many popular consumer products, such as computer printers and ink cartridges and video game consoles and the games themselves.

—Laura Lambert

Further Reading

Books
Adams, Russell B. *King C. Gillette: The Man and His Wonderful Shaving Device*. Boston: Little, Brown, 1978.

Dowling, Tim. *King Camp Gillette 1855–1932: Inventor of Disposable Culture*. London: Faber and Faber, 2001.

McKibben, Gordon. *Cutting Edge: Gillette's Journey to Global Leadership*. Boston: Harvard Business School Press, 1997.

Web sites
Lemelson-MIT Program: Inventor of the Week—King C. Gillette
   A biography and portrait of Gillette.
   http://web.mit.edu/invent/iow/gillette.html

Safety Razors and Shaving Collectibles
   A hobby site featuring photographs and technical information.
   http://www.safetyrazors.net/

See also: Household Inventions.

CHARLES GINSBURG

Developer of the videotape recorder
1920–1992

Charles Ginsburg provided the leadership vital to developing the first videotape recorder. He assembled the team of development engineers responsible for creating the recorder and pushed them to create the groundbreaking new device.

EARLY YEARS

Charles Ginsburg was born in San Francisco, California, on July 27, 1920. When he was four years old, he was diagnosed with diabetes, a condition in which the body fails to produce the hormone insulin. Fortunately for Ginsburg, insulin had recently been made available to treat this potentially fatal disease, so he was able to keep his condition under control.

As a young man, Ginsburg was interested in science but was uncertain about what area he wanted to study. He entered the University of California–Berkeley as a premedical student, then transferred two years later to the University of California–Davis to study animal husbandry. Financial difficulties combined with an inability to focus prompted Ginsburg to drop out of college in 1940.

Ginsburg then worked at a variety of jobs, including as a sound technician for a San Francisco recording company in 1942. The next year Ginsburg worked as an engineer at a radio broadcasting company, a job he held until 1947. While working, Ginsburg returned to college to study engineering, earning a bachelor's degree from San Jose State University in 1948. A year before he graduated, Ginsburg switched jobs again, working as a transmitter engineer at a San Francisco radio station.

MAGNETIC TAPE

Undated portrait of Charles Ginsburg.

Ginsburg was eager to move into what was then the new field of television. He mentioned this to a neighbor who worked for Ampex Electric and Manufacturing Corp., a company in Redwood City, California. Ampex, founded in 1944 at the height of World War II, had been a military contractor, but with the end of the war it had moved into a new business: magnetic tape recordings.

Magnetic tape had been perfected for audio recording by German scientists during the war. After the war ended, American soldiers brought the technology back to the

Backstage at a comedy program, Caesar's Hour, in the 1950s, when all television shows were broadcast live.

United States, where it caught on in the radio industry because it was cheaper and made recordings of higher quality than any other technology then available.

The popularity of magnetic tape in radio broadcasting suggested to Ampex's management that it might also be used in television. At that time, the lack of an inexpensive, high-quality method to record video was a major problem for television studios. Most shows were transmitted live—an approach that had serious drawbacks because any mistakes or inappropriate remarks were also immediately broadcast. If a network wanted to record a show to televise later—for viewers living on the West Coast, for example—the show had to be filmed and the film quickly developed and then broadcast. This process was both difficult and very expensive—film is costly and cannot be reused. The need for videotape recordings was so apparent that Ampex was competing with a number of other, larger corporations, among them the giant Radio Corporation of America, to develop the technology.

Recording video, however, presented a serious problem: a video signal contained much more information than an audio signal, and no one knew how to fit all the information onto magnetic tape. Magnetic tape was made by coating thin plastic tape with iron oxide powder, which became permanently magnetized when exposed to a magnetic field. To make a recording, a recording head containing an electromagnet was passed over the tape. The strength of the magnetic field varied depending on what was being recorded, and the tape was affected by those variations. During playback, the variations in the tape's magnetism were translated into sounds or images.

> The development of the first practical television videotape recorder did not flow from divine inspiration or a miraculous breakthrough onto the road to success. The first video recorder was the end product of over four years of hard, and at times, inspired work by a team of individuals.
>
> —Charles Ginsburg

In audio recordings, the information is recorded in one long, straight line running along the tape. When engineers first tried to make videotape, they also recorded the information in a long, straight line along the tape. However, because the video signal contained so much more information, recording a very short video clip required an enormous amount of tape—the length of tape that could record a half hour of music could record only a minute of video. Because the tape had to be so long, the video clip could be recorded or played only at such high speeds that, should the tape break, people could be seriously injured.

Ampex's head engineer, Walter Selsted, had seen what he thought might be a solution—an experimental videotape recorder with the recording and playing heads (the interface between the tape and the recorder) put on a small wheel that spun. The spinning heads permitted information to be recorded in countless small strips that ran across the tape rather than in one long strip along the tape. The tape did not have to be played at a high speed, because the heads were themselves moving quickly.

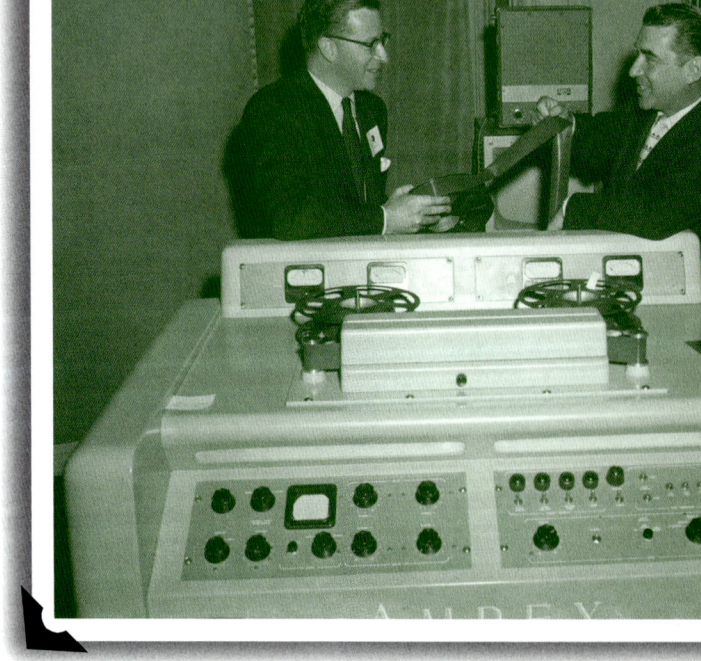

William Lodge (left), vice president of CBS, and George Long (right), president of the Ampex Corporation, during a demonstration of Ampex's new video recorder in 1954.

GINSBURG COMES TO AMPEX

Ampex's management decided to develop a videotape recorder with revolving heads. Ginsburg was recommended for the post of lead engineer for the program, and in January 1952, he went to work for Ampex.

Ampex, however, was working on several other projects at the time, and the videotape recorder was a fairly low priority. In May 1952, the program was suspended for three months to allow Ginsburg to help on a project to develop another recording technology.

While working on that project, Ginsburg met Ray Milton Dolby, a college freshman who worked part-time at Ampex. Dolby, who would later develop the noise-reduction technology that carries his name, had heard about the videotape recorder project and was very excited about it. Ginsburg was impressed by Dolby's ability and enthusiasm. When the effort to build a videotape recorder was reinstituted in August, Dolby was made a part of the team.

SETBACKS

In November 1952, Ginsburg and Dolby built a prototype videotape recorder. The quality of the recorded image was very poor, however, and Ampex's management was unimpressed.

The project received another blow in the spring of 1953, when Dolby, who had dropped out of school to work on the videotape recorder, was drafted into the U.S. Army. It seemed unlikely that the project would ever be a success. In June, Ampex suspended the videotape recorder project once more.

However, Ginsburg continued to work on the videotape recorder in secret. He was helped by another engineer, Charles Anderson, who joined Ampex in the spring of 1954. In August 1954, Ginsburg and Anderson made a presentation of an improved videotape

Illustrations from the 1958 patent for a magnetic tape recording and reproducing system created by Ginsburg and his team at Ampex.

recorder to Ampex's management. The quality of the images was much better, suggesting that perhaps the recorder could be successful. Ampex decided to restart the project once again.

SUCCESS

The recorder was far from perfect. As the heads spun, they tended to create white lines across the video images. Ginsburg quickly went about assembling an engineering team to overcome this and other problems; the team included Anderson, Shelby Henderson, Fred Pfost, and Alex Maxey; when Dolby returned from the army in January 1955, the team was complete. The six men threw themselves into work on the videotape recorder, developing new recording and playing heads, altering the design of the tape guide, and making many other improvements.

Ginsburg's team took a little over a year to perfect the video recorder, which was called the Mark IV. The device was made public in April 1956 (see box, The Unveiling of the Mark IV). The first television broadcast using videotape was made the following November.

FROM VTR TO VCR

A referee checks the video replay during a football game at Chicago's Soldier Field in 2005.

Ginsburg remained with Ampex until his retirement in 1986. Four years later, he was inducted into the National Inventors Hall of Fame. He moved to Eugene, Oregon, where he died in 1992 at the age of 71. In 2005, he, Dolby, Maxey, Anderson, Pfost, and Henderson were awarded the first lifetime achievement technology and engineering Emmy award by the National Television Academy for their contributions to the field.

Ginsburg's contribution was far-reaching. Videotape recording (VTR) gave television studios a practical means of recording shows; taped shows eventually surpassed live broadcasts in popularity. In addition, VTR made tele-

TIME LINE

1920	1948	1952	1956	1986	1992
Charles Ginsburg born in San Francisco, California.	Ginsburg earns a bachelor's in engineering from San Jose State University.	Ginsburg begins to develop his videotape recorder at Ampex.	Ginsburg's team unveils the Mark IV video recorder.	Ginsburg retires from Ampex.	Ginsburg dies.

vision broadcasts more professional, eliminating the mistakes and dead airtime that had previously plagued the industry. In the 1970s, videotape recording technology was miniaturized, creating the videocassette recorder (VCR), the first device ordinary consumers could use to record and watch video at home.

Home video would threaten the entertainment industry by reducing the number of people who watched broadcast television, but would also

The Unveiling of the Mark IV

By early 1956, Ampex's managers were so impressed by the quality of the video recordings Ginsburg produced that they decided to give a surprise demonstration of the recorder at a meeting in Chicago in mid-April. In attendance were more than two hundred managers of television stations affiliated with CBS Broadcasting.

The videotape recorder, called the Mark IV, was transported to Chicago without incident. The day before the meeting, however, the Mark IV was demonstrated to a CBS official, who complained that the image quality was not quite good enough. The engineers realized that they needed better-quality tape and made a last-minute phone call to the tape manufacturer, 3M Company, which flew out an experimental new videotape in time for the demonstration.

The demonstration was an enormous success and was met with cheers and applause. Ampex had expected to sell perhaps five of the Mark IVs, which cost $45,000; the company received orders for 82 machines by the end of the month.

provide new opportunities to make money as ordinary individuals bought videos of their favorite shows and movies. The work of Ginsburg and his team eventually altered nearly every aspect of the television industry.

—Mary Sisson

Further Reading

Book
Abramson, Albert. *The History of Television, 1942 to 2000*. Jefferson, NC: McFarland, 2003.

Web sites
Hall of Fame/Inventor Profile
  A sketch of Ginsburg from the National Inventors Hall of Fame Foundation.
  http://www.invent.org/hall_of_fame/66.html
The Race to Video
  A history of the development of video recording from *Invention and Technology* magazine.
  http://www.americanheritage.com/articles/magazine/it/1994/2/1994_2_52.shtml

See also: Communications; Entertainment.

ROBERT H. GODDARD

Inventor of rockets

1882–1945

New inventions are often greeted with surprise and delight, but they can also be met with disbelief and ridicule. When Robert Goddard suggested sending rockets into space, people thought he was crazy, and his work was largely ignored in the United States during his lifetime. Within a quarter century of his death, however, astronauts had set foot on the moon. Today, Goddard is widely recognized as the father of the space rocket.

EARLY YEARS

Robert Hutchings Goddard was born on October 5, 1882, in Worcester, Massachusetts. Both his parents were only children and Goddard, too, became an only child in 1894 when his brother Richard died soon after birth. Goddard's parents were very protective of him, and he frequently missed school for long periods because of illness. A curious child, Goddard made the most of these times, losing himself in books about physics and inventions.

He began dreaming of space flight around the age of 16, in 1898, when he read the science fiction novel *War of the Worlds* by H. G. Wells, which had been serialized in the *Boston Post*. The following year, the tall, gangly Goddard climbed a cherry tree in his family's garden to prune some branches. Perched high above the ground, as he later remembered, "I imagined how wonderful it would be to make some device which had even the possibility of ascending to Mars, and how it would look on a small scale, if sent up from the meadow at my feet." This idea stayed with him all his life.

Goddard attended South High School in Worcester from 1901 to 1904, then Worcester Polytechnic Institute from 1904 to 1908. There, his interest in technology grew—and he conceived of his first invention. When he was asked to write an essay imagining how people might be traveling 50 years in the future, he dreamed up a futuristic railroad system in which people shot back and forth inside metal tubes pulled by vacuum pumps and electromagnets. During his time at the institute, he carried out some of his first experiments. Goddard gained particular notoriety when he fired a test rocket in the basement of the school, causing a loud, smoky explosion.

Robert H. Goddard, photographed around 1935.

EXTREME ALTITUDES

Goddard's serious rocket work began when he moved to Clark University in Worcester, where he studied for his doctorate between 1908 and 1911. For a couple of years he worked at Princeton University in New Jersey before returning to Clark as a physics instructor in 1914.

He was granted his first two patents in that year. One was an idea for powering rockets with liquid fuels instead of the solids—gunpowder, for example—that were used in missiles and fireworks. Goddard realized that liquid fuels had advantages, not least because they could be switched on and off to steer a rocket into orbit. The second patent was an idea for a multistage rocket: one that would enter space as a series of connected parts or stages, each of which would push the rocket higher into space, then fall away. These two ideas—liquid fuel and multiple stages—are used in virtually every modern rocket launch.

> It is difficult to say what is impossible, for the dream of yesterday is the hope of today and the reality of tomorrow.
> —Robert Goddard

Goddard financed his early research himself, but as his ideas grew more ambitious, he needed more money. In 1916, he wrote to the Smithsonian Institution requesting help, attaching a 70-page report of his experiments, *A Method of Reaching Extreme Altitudes*. Impressed, the Smithsonian awarded him a $5,000 grant. Goddard's mother was delighted: "Think of it! You send the Government some typewritten sheets and some pictures, and they send you $1,000, and tell you they are going to send four more." In late 1919, the Smithsonian published Goddard's report as a booklet. Although most of it was given over to mathematical theory, Goddard used a few pages at the end to look into the possibility of travel to an "infinite altitude" and suggested sending a rocket to the moon.

The press latched onto this suggestion and ridiculed his ideas. A story on the front page of the *New York Times* on January 12, 1920, called his ideas "absurd" and suggested that Goddard lacked a grasp of the most basic physics "ladled out daily in high schools." How, the editors asked, could a rocket work in space, where there was no atmosphere for it to push against? Goddard had actually proved that rockets could work in the vacuum of space five years earlier with an ingenious experiment. He mounted a revolver on a pivot inside a vacuum chamber and fired it remotely. As the bullet left the gun, its momentum made the revolver recoil and spin around backward. Goddard figured that a rocket engine would work like the bullet, pushing the rocket in the opposite direction from which it fired. However, this proof did not satisfy the critics. Making matters worse, Goddard started receiving letters from people eager to come with him on his trip to the moon. One of the most famous movie stars of that era, Mary Pickford, got her agent to cable

Goddard and his liquid fuel rocket, photographed before its first flight on March 16, 1926.

Goddard this request: "WOULD BE GRATEFUL FOR OPPORTUNITY TO SEND MESSAGE TO MOON FROM MARY PICKFORD ON YOUR TORPEDO ROCKET WHEN IT STARTS."

The public ridicule only made Goddard more determined to continue his experiments. In the early 1920s, he worked for the U.S. government at Indian Head, Maryland, designing rocket-based weapons powered by solid fuels, but the idea still looming in his mind was to build a space rocket powered by liquids. In 1923, he designed a rocket engine that used gasoline (a liquid fuel) and liquid oxygen to burn it. The next year, on June 21, he married Esther Christine Kisk, then secretary to the president of Clark University, who became a great supporter of Goddard's work. Back at Clark University, where he had now been promoted to director of the physical laboratories, Goddard tested his liquid-fuel engine for the first time in 1925. To his delight, it produced enough thrust (upward force) to lift its own weight off the ground.

Throughout 1925, Goddard tried many different engine designs, tinkering with the two pumps that drove the liquid oxygen and gasoline from the tanks that stored them to the engine itself. By the fall of that year, he had greatly simplified the design so that it used only one pump. This achievement was a major step forward. A few months later, he had a working liquid-fuel rocket that he was ready to test. On March 16, 1926, he successfully launched it from his Aunt Effie's farm near Auburn, Massachusetts. Although the flight lasted just 2.5 seconds and the rocket flew only 40 feet (12 m) into the air, this first flight was as important as the Wright brothers' pioneering flight in 1903 (see box, How Goddard's Rocket Worked).

DESIGNING A ROCKET

Before rockets could carry people into space, Goddard knew he would need to test conditions in earth's atmosphere to ensure that such flight was safe. He began to send up rockets that could take scientific measurements. On July 17, 1929, still working at his aunt's farm, he launched the first rocket with instruments on board: a barometer to measure pressure, a thermometer to check temperature, and a small camera. Although the test was successful, some of the terrified local people saw the rocket's flame roaring through the sky and assumed a plane was about to crash-land. They called the fire department, ambulances, and the police. The fire chief banned any more flights from Aunt Effie's farm, and Goddard was forced to move his test site to an army artillery range.

The episode brought more ridicule from the press: one Worcester paper carried the headline: "MOON ROCKET MISSES TARGET BY 238,799$^{1}/_{2}$ MILES." However, a later story about Goddard's work caught the attention of Charles Lindbergh (1902–1974), who had gained worldwide fame in 1927 by making the first solo nonstop airplane flight across the Atlantic. When the two men met in November 1929, Goddard explained that he needed more money to continue his work; Lindbergh, who believed that Goddard was a true visionary, decided to help. The following year, Lindbergh won a grant from the Guggenheim Foundation on Goddard's behalf. Thanks to Lindbergh, the Guggenheims would donate nearly $200,000 to fund Goddard's research from 1930 through 1942.

With his finances at last secure, Goddard took an extended leave of absence from Clark University and relocated with his wife to Mescalero Ranch at Roswell, New Mexico, to carry out his experiments. From July 1930 until the mid-1940s, he developed bigger and better rockets and made many advances that would later prove invaluable to the U.S. space program. In 1932, he pioneered a way of using gyroscopes (heavy

How Goddard's Rocket Worked

"One small step for man . . ."

Many people think the space age began on July 20, 1969, when astronaut Neil Armstrong (1930–) took "one small step for man" across the dusty surface of the moon. Others think April 12, 1961, was a more important date, when Russian cosmonaut Yury Gagarin (1934–1968) became the first human to orbit the earth in a rocket. The age of space travel can actually make strong claims to have begun a half century earlier on March 16, 1926. On that day, Robert Goddard launched the first rocket powered by liquid fuel from his Aunt Effie's cabbage patch on a farm in Auburn, Massachusetts. That experiment made possible every space mission since.

Goddard's rocket worked like a large and complex firework—with a few crucial differences. In a firework, solid explosive chemicals burn in the case using oxygen from the air to produce a jet of exhaust gas. The force of a firework's exhaust flying backward produces an opposite force, thrust, that propels the firework forward. Unlike earth, space has no atmosphere and no oxygen. In addition to carrying fuel, a rocket must carry its own supply of oxygen to make the fuel burn.

One of Goddard's greatest discoveries was that liquid fuel works better in a rocket than the solid fuel used in a firework. Liquid fuel makes a rocket easier to control; however, it requires a system of pipes and pumps to move it from the tank, where it is stored under high pressure, to the rocket motor, where it burns. Although Goddard's rocket functioned like a firework, it looked more like plumbing made from steel and aluminum tubing. A modern space rocket has a similar (though much more elaborate) collection of pipes and pumps concealed inside its sleek, streamlined case. Goddard left the raw mechanism of his own rocket open to the air to save weight so the rocket could climb higher.

Launching Goddard's rocket was a risky operation. Poking out of the top of the combustion chamber was a fuse that his assistant had to light by hand. Once the fuse was burning, Goddard switched on the oxygen tank with a long wooden stick. Then he turned on a pipe from an oxygen cylinder outside the rocket to pump in the gasoline. When the gasoline and oxygen entered the combustion chamber, they ignited and the rocket took off. This method of ignition was far from a modern spacecraft's, launched by computer-controlled countdown.

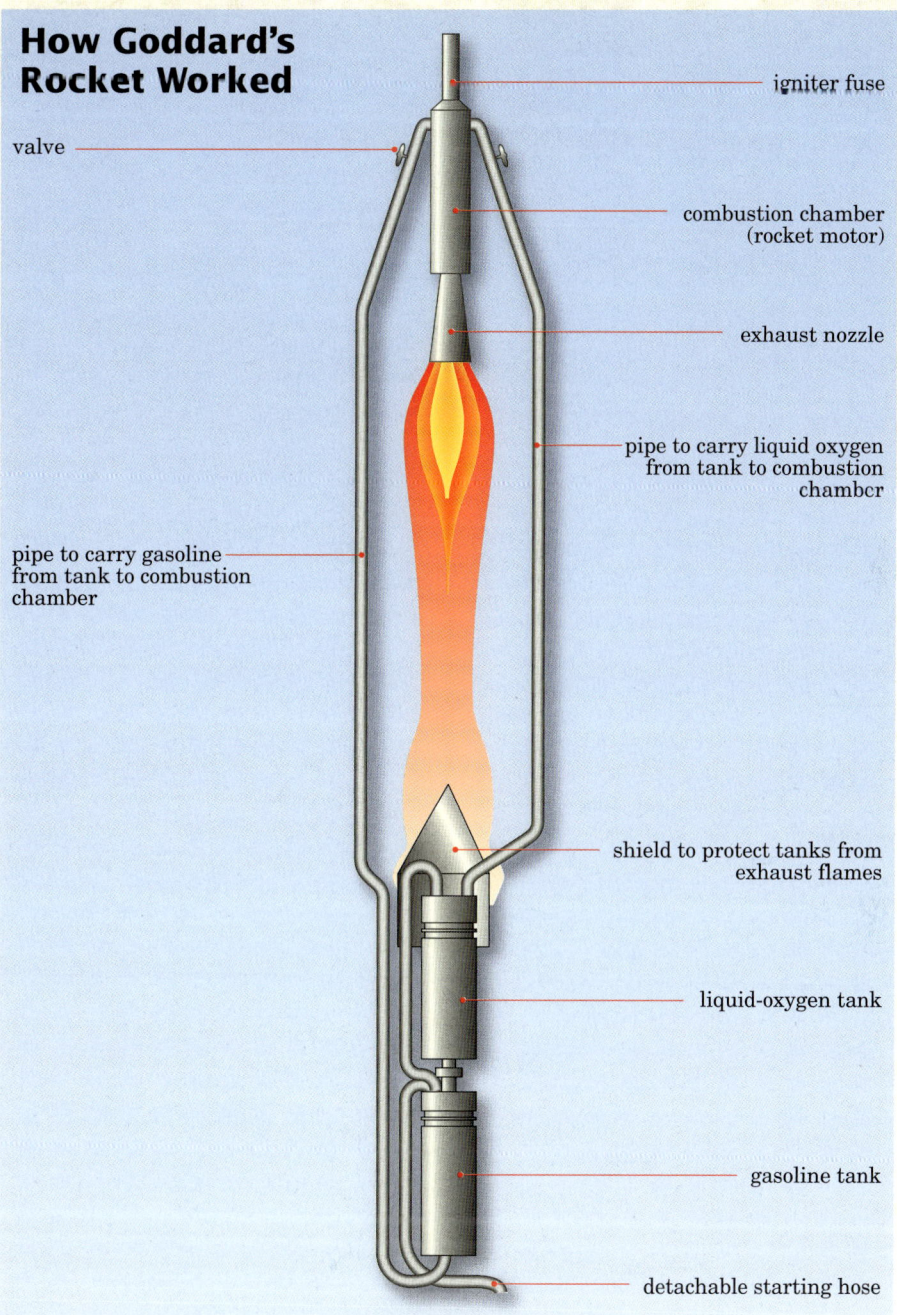

Goddard's rocket had four main parts. At the top was the combustion chamber, a metal cylinder inside which the liquid fuel (gasoline) mixed and burned with liquid oxygen (oxygen stored under very high pressure so it turns from a gas to a liquid). There was a flared exhaust nozzle directly underneath the combustion chamber from which the hot exhaust gases could escape. Three or four feet beneath the exhaust nozzle was a large, upside-down V-shaped pipe system running to a gray, missile-shaped section at the bottom. This contained two tanks, one for the liquid oxygen and the other for the gasoline, and various control valves and mechanisms.

At Goddard's launch tower in 1935 are (left to right) Albert Kisk (a machinist), Harry F. Guggenheim, Goddard, Charles A. Lindbergh, and N.T. Ljungquist (also a machinist).

spinning wheels) to give a rocket more stability in flight. The same year, he figured out how to use vanes (or fins) at the base of a rocket to control and steer it. Some of his greatest achievements came later that decade. In 1935, he fired a rocket faster than the speed of sound, which is approximately 700 miles per hour (1,127 km/h). Two years later, he fired another rocket to a record-breaking height of 1.7 miles (2.7 km).

During his time in New Mexico, Goddard lived as a secretive inventor, fearful that others would achieve his goals first. He was confident about his ideas. Although he had many disappointments, they never bothered him: he simply changed tack and started again. Working in isolation, however, was not an entirely good decision. Goddard was determined to build better rockets, but he trusted only a few people and refused to work with others who might have helped him. He had to promote his work to win the funds he needed, yet he shunned attention, fearing continued ridicule from the press. While his supporters, including Lindbergh and the Guggenheims, pressed him for results, he worked very slowly, secretively, and meticulously. He wanted to succeed, but only on his own terms. Ironically, Goddard might have built the high-altitude rocket he dreamed of if he had been willing to collaborate with others.

Goddard saw a new opportunity to put his work to use when the world plunged into World War II (1939–1945). In May 1940, he offered his lifetime of work—all his inventions, his rocket data, and his facilities at Roswell—to the U.S. armed forces. They turned him down, but eventually decided they could use his expertise in other ways. So, in 1942 he moved to Annapolis, Maryland, to help develop jet-powered seaplanes for the U.S. Navy. In the final two years of the war, Goddard, now in his early sixties, became a consultant engineer for the Curtiss-Wright airplane company. He experienced some small recognition of his work in June 1945, when Clark University made him an honorary doctor of science, but Goddard was by then seriously ill with throat cancer. He died only a few weeks later, on August 10, in Baltimore, Maryland, and was buried in Hope Cemetery at his family hometown of Worcester.

ROCKETS AFTER GODDARD

Just as Goddard's work was being brushed aside in the United States, the Germans, led by brilliant engineer Wernher von Braun (1912–1977), had seen the true potential of rockets. Ironically, the Germans had

How the Chinese Invented Rockets

When it came to rockets, Robert Hutchings Goddard was years ahead of his time—but Chinese inventors were hundreds of years ahead of him. Rockets first became possible when Chinese chemists invented gunpowder around 700–800 CE; extant evidence shows that the Chinese used simple rocket missiles in a siege against the Mongols in 1232 CE. These rockets, which looked and behaved just like modern fireworks, "arrows of flying fire," were packets of explosive chemicals tied onto long sticks: the explosives provided a rush of power that blew the rockets upward, and the sticks made them travel roughly in a straight line so they could be aimed at a target.

According to legend, a Chinese inventor also made a brave attempt at space flight about 250 years later. Around 1500, Wan Hu had a large rocket chair built; he had 47 "arrows of flying fire" attached to the chair and, on the day he chose for its launch, he climbed into it wearing ceremonial robes. Forty-seven officials reputedly lit the 47 rockets and stood well back. There was an enormous explosion and a large cloud of smoke, and Wan Hu was never seen again. Whether he flew into space or was blown to pieces, no one knows. However, NASA recognized his importance as a space pioneer when it named the Wan-Hoo crater on the moon for him.

The Apollo 11 launch on July 16, 1969.

recognized Goddard's genius when they obtained copies of his Smithsonian report, *A Method of Reaching Extreme Altitudes*. During World War II, they used Goddard's ideas to develop powerful rocket missiles known as the V-1 and V-2. These rockets were fired by the thousands on countries such as Britain, with devastating effect. When the war ended, von Braun and many of his colleagues in the German rocket division surrendered to Allied forces and moved to the United States, while some of the rest went to the Soviet Union. These scientists formed the core of the rival U.S. and Soviet space programs, which ultimately realized Goddard's dream of firing rockets into space in the late 1950s.

The importance of Goddard's work was finally recognized when the United States began to develop a space program. In his lifetime, Goddard had proved that rockets could fly in the vacuum of space; flown rockets faster than sound; designed rocket engines, liquidfuel pumps, and

TIME LINE

1882	1911	1914	1916	1923
Robert Hutchings Goddard born in Worcester, Massachusetts.	Goddard earns doctorate in physics from Clark University.	Goddard begins teaching physics at Clark.	Goddard receives grant from Smithsonian Institution to fund research.	Goddard designs liquid-fuel engine.

steering mechanisms; launched the first high-altitude rockets; used gyroscopes and fins to guide and control a rocket after launch; and sent the first instruments into the sky inside a rocket. After his death, the U.S. Patent Office awarded 214 different patents for various aspects of rocket design to his estate, and the U.S. government paid $1 million to use them. Since then, virtually every space rocket launched has been based on Goddard's ideas. In 1959, the American space agency, NASA, formally recognized Goddard's work by naming its Goddard Space Flight Center for him. The day after the moon landing in July 1969, the *New York Times* published a correction to its 1920 editorial, noting: "It is now definitely established that a rocket can function in a vacuum as well as in an atmosphere. The *Times* regrets the error."

Goddard did not live to see people traveling into space or setting foot on the moon. Doubtless he would have been delighted to have seen the

A Titan/Centaur rocket launches the Viking 1 probe on its 11-month journey to Mars in 1976.

TIME LINE (continued)

1925	1926	1930–1942	1945
Goddard tests his liquid-fuel engine.	Goddard launches the first liquid-fuel rocket.	Goddard moves to New Mexico to conduct experiments.	Goddard dies.

U.S. Mariner and Viking probes making the first voyages toward Mars in the 1960s and 1970s, just as he had dreamed in his cherry tree in 1898. None of these developments would have come as any surprise to Goddard, who was certain that people would one day fly into space. He endured ridicule from the public and the press and the military's lack of interest—a price many inventors pay for being decades ahead of their time.

—Chris Woodford

Further Reading

Books

Clary, David. *Rocket Man: Robert H. Goddard and the Birth of the Space Age*. New York: Theia, 2003.

Patchett, Kaye. Robert *Goddard: Rocket Pioneer*. Farmington Hills, MI: Blackbirch, 2005.

Streissguth, Tom. *Rocket Man: The Story of Robert Goddard*. Minneapolis, MN: Carolrhoda Books, 1995.

Web sites

Brief History of Rockets
   A chronology of rocketry compiled by NASA's space scientists.
   http://quest.arc.nasa.gov/space/teachers/rockets/history.html

Dr. Robert H. Goddard Home Page
   A collection of Goddard's papers at Clark University.
   http://www.clarku.edu/offices/library/archives/Goddard.htm

Smithsonian National Air and Space Museum
   Many exhibits about the history of flight, including space rockets.
   http://www.nasm.si.edu/

See also: Military and Weaponry; Transportation.

PETER GOLDMARK

Inventor of modern color television
and the long-playing record

1906–1977

Although an inventor may devise elegant and creative solutions to pressing problems, the widespread adoption of an invention is often the result of factors lying outside the inventor's control. This was particularly true for Peter Goldmark. During Goldmark's long career at what is now the CBS network, he saw some of his inventions—for example, the long-playing record— alter entire industries. Other inventions, however, like his method for producing color television broadcasts or his early form of home video, languished because of lack of industry and regulatory support.

EARLY YEARS

Peter Goldmark was born in 1906 in Budapest, Hungary. His parents divorced when he was eight. His mother remarried a banker about a year later, and Goldmark had a relatively comfortable upbringing, despite continuing outbreaks of political violence in Hungary. Because of the unrest, the family decided to move to Vienna, Austria, when Goldmark was a teenager.

Goldmark's extended family was very musical—his great-uncle was the composer Karl Goldmark (1830–1915)—but Goldmark was more interested in science and engineering. He attended the Berlin Technische Hochschule for college but disliked the school's electrical engineering program. After a year, he moved to the Physical Institute at Vienna, where he eventually earned a bachelor's degree and a doctorate in physics.

In 1926, while studying for his doctorate, Goldmark heard from a friend about a do-it-yourself kit, made by a British company, that would allow the buyer to experience a new technology: television. Goldmark sent away for the kit, assembled the machine, and waited until midnight, when England's BBC network broadcast a video signal over the radio airwaves.

The broadcast featured a dancer. The screen was only about one inch high (2.54 cm) and half an inch (1.27 cm) wide, and the quality was not very good. Still, Goldmark later wrote, "when she was there before us and reasonably clear, what a sight! . . . Even in the 1920s, television had the quality of wonder to it, at least to me."

COLUMBIA BROADCASTING SYSTEM

Peter Goldmark, photographed in 1968.

Goldmark became intrigued by television. First, he devised a method to make the viewing screen larger, patenting it in Austria (a somewhat challenging task, as the patent examiner had never heard of television). England seemed to be the center for this nascent technology, so Goldmark moved there, taking a job in the television department of Pye Radio, Ltd. Television at that point was not a money-making business, and with the onset of the economic depression of the 1930s, Pye decided it could no longer afford to develop the novel technology. Consequently, Goldmark was laid off and he returned to Vienna.

Goldmark then decided to immigrate to the United States, where the interest in developing television was greater. He arrived in New York City in 1933 and worked for a radio manufacturer as an engineer while looking for a job in television. In late 1935 he received a job offer from the Columbia Broadcasting System (CBS) to head its television research effort.

Like many companies interested in television at the time, CBS was primarily a radio broadcast network. Television was still experimental—CBS would not make its first television broadcasts until 1939—and the business potential was unclear. As an employee of CBS, Goldmark contributed to the development of the engineering that made reliable television broadcasts possible.

In March 1940, Goldmark saw the color movie *Gone with the Wind*. Color movies were rare at the time, and the experience convinced

How Goldmark's Color Television Worked

When Goldmark decided to create color television, he needed a system that would transmit all the information required to create full color. However, the available spectrum of airwaves was limited, so he could not transmit information for every single color.

Instead, Goldmark developed the field-sequential system. In this system, a show would be shot through three filters: one red, one blue, and one green. The three filters were placed on a rotating disk located behind the camera lens.

The rotation of the filters was synchronized with an electron-scanning beam, which translated the image into a signal. As a result, first a red image, then a blue image, and last a green image was transmitted to television sets in homes.

The images were transmitted at separate intervals. The viewer at home, however, would see full color because the images were rotated so quickly that the viewer's brain would blend the colors together, creating a full-color image.

Goldmark that television should be broadcast in color, rather than black-and-white. He immediately went to work on the problem, and within months he had developed a working color television system. Goldmark's color television, however, quickly ran into trouble. At the time, broadcasters and the Federal Communications Commission (FCC), which regulates the airwaves in the United States, were trying to settle on industry-wide standards. Introducing a new kind of television might have the effect of delaying developments in broadcasting.

Before any resolution could be reached on the matter of color television, the United States was drawn into World War II. Further development of color television had to wait until the war concluded.

INVENTING THE LP

Goldmark, who had become a U.S. citizen in 1937, spent the war developing systems to jam German radar. After the war ended, he returned to his job at CBS in New York City, expecting to continue work on color television. In the fall of 1945, Goldmark attended a party that would inspire him to develop a new technology: the long-playing record.

The hosts of the party had a new recording of Johannes Brahms's Second Piano Concerto, played by the famous pianist Vladimir Horowitz. At the time, a 12-inch (30-cm)-diameter record held only about five or six minutes of music. Every few minutes, the music would stop, and someone would have to change the record. For Goldmark, an avid lover of classical music, the interruptions were an abomination: "I knew right there and then I had to stop that sort of thing," he later recalled.

What he would later call his "sincere hatred for

In 1948, Peter Goldmark poses next to a tower of 78-rpm records; the 33$\frac{1}{3}$-rpm records he holds under his arm contain an equivalent amount of music.

the phonograph" was not the result only of short playing time. Records were then made of shellac and played with heavy steel needles. To keep the needles from wearing out the records, manufacturers put a hard material in the grooves to protect them. However, this substance was rough, resulting in a fuzzy background noise that listeners often found extremely distracting.

As turntables of the day played 12-inch records, Goldmark decided to keep that size for his invention. To make his record play longer, he slowed the playing speed from 78 to $33^1/_3$ revolutions per minute. He then made the grooves in the record, which contained the auditory information, narrower to fit more grooves on a disk. Such narrow "microgrooves" could not be played with a heavy steel needle; accordingly, Goldmark developed thinner needles and lighter record player arms.

Goldmark also got rid of shellac, pressing his record out of a much sturdier substance—Vinylite. These vinyl records were more expensive to produce, but Goldmark argued that the production costs would be lower, as fewer records would be needed for the same amount of music.

In 1948, CBS unveiled the long-playing record, which could play about 45 minutes of music on each side. Shortly before its public debut, CBS executives tried unsuccessfully to find a catchy name for the new technology. "I guess the LP isn't going to have a name after all," said Goldmark, using his laboratory's shorthand term for "long player." An executive heard him, and the new record was dubbed the LP.

> As I look back, I think my contributions were, somewhat ironically, not so much in the invention itself or in innovation (a word I prefer because it means putting an invention to work), but in its gadfly impact on industry.
>
> —Peter Goldmark

COLOR CONTROVERSY

During the late 1940s Goldmark and CBS were still trying to get the FCC to approve Goldmark's color television system. They faced a formidable rival, however, in the Radio Corporation of America (RCA), another radio company turned television company that was creating its own color television system. Goldmark's system would require the owner of a black-and-white television to buy and install a special adaptor. It actually worked, whereas RCA's system was still under development.

In 1950, the FCC granted the first commercial license to broadcast color television to CBS. RCA sued, receiving an injunction against CBS that prevented it from broadcasting in color. CBS appealed to the U.S. Supreme Court in 1951 and won. However, the highest court's ruling upheld a lower court's ruling that no color televisions could be manufactured at the time; the Korean War was under way and cobalt, a material needed for color televisions, was on the list of critical materials needed for the war effort.

A cooking program on CBS, aired during the first week of color television broadcasts in July 1951.

RCA used the time it had been given wisely, undertaking a huge effort to perfect its color television system. In 1953, with the war over, the FCC revisited the question and ruled against CBS: RCA's color television would be the industry standard.

LATER YEARS

In 1958 CBS moved its research laboratory from New York City to Stamford, Connecticut, where Goldmark lived. The move resulted in part from a restructuring of the company: Goldmark's laboratory was now a stand-alone division that was free to make products for other companies and institutions as long as it turned a profit. As a result

TIME LINE

1906	1926	1933	1935	1940
Peter Goldmark born in Budapest, Hungary.	Goldmark experiences television for the first time.	Goldmark moves to New York City.	Goldmark begins work at CBS.	Goldmark develops a working color television system.

The General Electric 15CL100, the company's first color television set, was released in 1954.

of this change, Goldmark became involved in the U.S. space program in the 1960s, helping develop technology that enabled satellites to take and transmit high-quality pictures.

Goldmark also developed a technology that he called electronic video recording (EVR) a predecessor to the home videocassette player. EVR allowed users to play prerecorded videos on their televisions, although they could not record broadcast programs. The EVR languished, however, because of what Goldmark viewed as a lack of support from CBS, which was concerned that the system would cut into broadcast ratings.

In 1971, Goldmark turned 65, which was the mandatory age of retirement at CBS. He was offered an opportunity to stay on as a consultant but turned it down because of his experience with the EVR; he then formed his own company. Goldmark was presented the National Medal of Science in November 1977; two weeks later, he died in a car crash in Westchester County, New York, at the age of 71.

GOLDMARK'S INFLUENCE

The LPs Goldmark invented would prove to be a highly popular musical format, being sidelined only by the advent of compact discs in the late 1980s. The LPs also gave rise to the high-fi(delity) stereo move-

TIME LINE

1945	1948	1971	1977
Goldmark begins to develop the long-playing record (LP).	CBS unveils the LP.	Goldmark retires from CBS.	Goldmark dies.

ment popularized in the 1970s that is still going strong. Because LPs produced high-quality sound, listeners wanted equipment and speakers that could accurately reproduce this sound.

Goldmark clashed with CBS management because he often had grand designs for the future use of his inventions; CBS wanted products it could sell right away. Although most of Goldmark's inventions were never widely adapted, they had significant effects on the television industry. Without the prospect of losing the potential market in color televisions to CBS, RCA might not have made an aggressive push to develop color television. Even EVR made an impression when it debuted: newspapers breathlessly covered its development, predicting a new future in home entertainment in which the consumer, not the broadcaster, chose what would appear on television. Although EVR never became popular, other technology would eventually bring Goldmark's vision to fruition.

—Mary Sisson

Further Reading

Books
Coleman, Mark. *Playback: From the Victrola to MP3, 100 Years of Music, Machines, and Money.* Cambridge, MA.: Da Capo, 2003.
Goldmark, Peter C. *Maverick Inventor: My Turbulent Years at CBS.* New York: Saturday Review, 1973.

Web sites
Color TV's 50th Anniversary
    NPR stories and videos marking half a century of color television.
    http://www.npr.org/templates/story/story.php?storyId=1789944
Early Television Museum
    Essays about and photographs of early television.
    http://www.earlytelevision.org/index.html
A History of Vinyl
    A BBC exhibit on the development of the record.
    http://www.bbc.co.uk/music/features/vinyl/

See also: Edison, Thomas; Entertainment; Farnsworth, Philo.

CHARLES GOODYEAR

Inventor of vulcanized rubber

1800–1860

Many inventions start out as good ideas. Nevertheless, they can take years to develop and just as long to turn a profit. Occasionally, however, inventors make unexpected breakthroughs that help them overcome obstacles they have struggled with for some time. A lucky break like this helped American inventor Charles Goodyear turn rubber into a much more versatile substance. The process he discovered, vulcanization, enabled rubber to become one of the most widely used materials of modern times.

EARLY YEARS

Charles Goodyear was born on December 29, 1800, in New Haven, Connecticut. Instead of attending school, he worked for his father, Amasa, a miller and farmer who also manufactured ivory buttons. At age 16, Charles left home and moved to Philadelphia, where he learned metalworking skills in a hardware business. When he returned to Connecticut, he went into business with his father manufacturing agricultural implements. In 1824, he married Clarissa Beecher, who would turn out to be his most loyal supporter.

The couple went back to Philadelphia two years later and opened a hardware store selling the implements the Goodyears had made in Connecticut. Everything went well for several years until Goodyear's health failed. Unable to work hard, and with some of his customers defaulting on their payments, Goodyear was plunged into debt. He soon realized he needed a new way of making money—some clever invention, perhaps—and turned his attention to rubber.

THE VALUE OF RUBBER

By the early 19th century, people had been using rubber in Europe and North America for around a century. Long before Columbus arrived in the Americas, the Aztecs are believed to have made rubber balls, as well as footwear, bottles, and a variety of other items.

During the 18th century, rubber spread to Europe, where it was soon being used in all sorts of new ways; covering cloth in liquid rubber was an effective way to make it waterproof, for example. Charles Macintosh (1766–1843) used this idea to develop his famous mackintosh raincoat in 1823. His partner, Thomas Hancock (1786–1865), developed a roller covered in spikes. This pickle machine, as it was called, was something

An undated portrait of Charles Goodyear.

like an automated mouth—when tiny strips of rubber were fed under the roller, it chewed them together to produce a lump of rubber suitable for making larger items. Despite such advances, rubber still presented a problem: when hot, it became smelly and sticky; when cold, it was hard and brittle. Rubber had terrific potential, but its variability greatly reduced its value.

A RUN OF BAD LUCK

Charles Goodyear turned his attention to this problem in the early 1830s. After reading everything he could find about rubber, he made some small rubber valves for life preservers and determined to sell them. When he took them to a retail store, however, the proprietor told him there was no market: no one wanted rubber goods because they melted in the hot weather.

Goodyear was determined to overcome this difficulty. Before he could, however, someone to whom he owed money had him arrested and thrown in jail. Goodyear asked his wife to bring him some rubber and a rolling pin and promptly started experimenting in his cell. The first thing he considered was the problem of stickiness. He thought he might overcome this by mixing rubber with a dry powder. When he was released, he bought some magnesia, which was similar to talcum powder, and worked it into the rubber, then boiled the mixture with quicklime and water. Convinced that he had cured the stickiness, he manufactured hundreds of rubber overshoes in his kitchen. When

A drawing depicts Goodyear demonstrating his vulcanization process (artist and date unknown).

Rubber

Rubber, which Charles Goodyear called gum elastic, earned its name in 1770. A British chemist, Joseph Priestley, was playing with some rubber when he found by accident that it would erase pencil marks if he rubbed it vigorously on the paper.

Rubber does not start off as a hard substance or even as a solid. It is manufactured from latex, a white liquid similar to sap, that can be drained from around 200 different plants by making a cut, or tap, in their bark. The best plant for rubber tapping is a species of tree, *Hevea brasiliensis*. As its name suggests, it originated in Brazil in South America.

Demand for rubber increased greatly in the late 19th century. In 1876, Henry Wickham, an explorer, collected *Hevea* seeds and smuggled them out of Brazil to London. They were then used to establish huge rubber plantations in countries that were then English colonies, including India, Ceylon (now Sri Lanka), Malaysia, and Singapore. Wickham's original seeds developed into the enormous Asian rubber industry that now produces almost all of the world's natural rubber.

By the early 20th century, rubber had become a vitally important military material (in everything from airplanes to gas masks). Synthetic rubbers really came into their own when an increasing demand for natural rubber pushed prices to astronomical levels in the 1920s; problems intensified when World War II began at the end of the following decade. Countries on both sides set some of their best chemists to the task of developing artificial rubbers from petroleum and other carbon-based chemicals. In the United States, Wallace Carothers and his team at the DuPont Company developed an artificial rubber, neoprene. Modern synthetic rubbers, such as SBR (styrene-butadiene rubber), are now used to make everything from car tires and sneakers to floor coverings and adhesives.

A rubber tree is tapped for latex in contemporary Malaysia.

> [My invention] may, therefore, be considered as one of those cases where the leading of the Creator providentially aids his creatures, by what are termed "accidents," to attain those things which are not attainable by the powers of reasoning he has conferred on them.
>
> —Charles Goodyear

summer came, however, they melted and turned into a soggy, smelly, and very sticky mass.

Goodyear knew there must be a way of making rubber more long-lasting. He continued his experiments. When his neighbors complained about the smell, he relocated his family to New York City and continued tinkering there in an attic. Sometime in 1836–1837, he discovered that he could make the surface of rubber smooth, dry, and harder by treating it with nitric acid.

On the strength of this discovery, he found a backer who advanced several thousand dollars to begin production of rubber goods. Soon Goodyear and his partner set up a factory on Staten Island and began making clothes, shoes, and other items. Then came the financial panic of 1837, which plunged the U.S. economy into chaos and many people into debt. Goodyear's partner lost his fortune and the factory went bankrupt. Goodyear was yet again back at the beginning. He and his family moved into the rubber factory and survived by eating fish they caught in the river.

Ever restless, Goodyear moved his family to Massachusetts. His family's situation began to improve when Goodyear won a contract to supply the U.S. government with mailbags made from his acid-treated rubber. He got as far as making 150 of them, but when he stored them in a warm room temporarily he found that, beneath the hard, acid-treated surface, the rubber had simply melted again. Goodyear knew he would never succeed until he could find a way of making rubber permanently durable. He came a step closer to solving this problem in 1838, when he met Nathaniel Hayward (1808–1865), operator of a factory in Woburn, Massachusetts. Hayward had recently discovered that rubber could be made less sticky by adding sulfur. Although his process was far from perfect, Goodyear sensed that there was something in it and paid Hayward for his patent rights.

A LUCKY BREAK

The breakthrough finally came in 1839. Goodyear was experimenting with some rubber on which he had sprinkled sulfur, an additive to make the rubber less sticky, when he dropped the rubber onto a hot stove by accident. He lifted the scorched rubber from the stove, and hung it to cool on the door frame. The next day, Goodyear found a much stronger and more robust material—what eventually came to be known as vulcanized rubber (see box, How Vulcanization Works).

Unfortunately, over the next few months his health deteriorated. His family was in a constant state of grinding poverty, living on food handouts from neighbors. Of the 12 children born to Goodyear and his wife, six died in infancy.

Rubber bands were patented in England in 1845 by Stephen Parry, a businessman who sold many early latex products.

During the next five years, Goodyear carried out many more experiments to discover the perfect way of treating rubber. He tried cooking it at different temperatures, for longer and shorter times, and with various chemicals. By 1844, he had a product that was good enough to patent. Instead of manufacturing rubber products himself, however, he planned to sell licenses, so other people would pay him royalties for the right to use his invention. In theory, this could have made him a fortune, but the deals Goodyear signed were bad ones. Companies paid him tiny royalties and then sold their rubber products at huge profits.

PATENT BATTLES

Goodyear's invention was a simple one, and other people started to copy it. Between 1844 and 1852, he fought 32 lawsuits against people who had tried to develop similar heat treatments. Finally, in 1852, he obtained a perma-

TIME LINE

1800	1817	Early 1830s	1836–1837	1839
Charles Goodyear born in New Haven, Connecticut.	Goodyear moves to Philadelphia to learn metalworking and hardware business.	Goodyear begins studying rubber.	Goodyear discovers that nitric acid can make rubber smooth, dry, and hard.	Goodyear discovers vulcanized rubber.

nent injunction, hoping to stop people from using his ideas—but it made little difference. That year, he realized his idea would be just as popular overseas as it was proving to be in the United States, so he traveled to Europe to try his luck there. At trade exhibitions in England and France, he built giant demonstration pavilions made entirely from rubber. The value of the new material was immediately apparent to everyone who saw the demonstrations.

Things did not go entirely as planned, however. When England's rubber pioneer, Thomas Hancock, saw a sample of Goodyear's scorched rubber, he managed to figure out how it was made and promptly devised a similar heat-treatment process of his own. Hancock called his process vulcanization, named for Vulcan, the Roman god of fire. The name was soon taken up to describe any treatment of rubber with heat and sulfur, including Goodyear's. Hancock rushed to protect his idea, filing an English patent several weeks before Goodyear managed to submit one of his own. When Goodyear began a lawsuit against him for patent infringement, Hancock offered to share half his royalties if Goodyear would drop the suit. Certain he would win in the courts, Goodyear refused—and then promptly lost his case.

Goodyear fared little better in France. In 1855, his French patent was canceled on a technicality, because American-produced rubber was already being imported to that country. This gave companies that were supposed to be paying Goodyear patent royalties a perfect excuse to stop. He suddenly found himself with a huge hotel bill to pay and no money, so the French police hauled him off to jail. Ironically, the same year he received the Cross of the Legion of Honor, in recognition of his work, from the French emperor Napoleon III.

LATER YEARS

By the time Goodyear returned to the United States, his patent was being widely infringed and it was virtually impossible for him to make any money from it. During the last few years of his life, he continued to refine his invention, but never really focused his mind on making money. Instead, he

TIME LINE

1844	1852	1855	1860	1898
Goodyear sells licenses for use of his rubber.	Goodyear travels to Europe to demonstrate rubber.	Goodyear's French patent cancelled.	Goodyear dies.	The Goodyear Tire & Rubber Company is founded.

The name Goodyear is strongly associated with tires, but the Goodyear tire company was named only in honor of the inventor and was not his company.

devoted himself to improving the process, experimenting with new treatments, and thinking up ever more uses for rubber. When he wrote his autobiography, *Gum-Elastic and Its Varieties*, between 1853 and 1855, he bound it in a rubber cover. Charles Goodyear died on July 1, 1860, in a room at the Fifth Avenue Hotel in New York City. The inventor of one of the world's most important materials was approximately $200,000 in debt.

The Goodyear Tire & Rubber Company, named in his honor, was founded in 1898—almost 40 years after his death. Although neither he nor his family made money from it, the company did at least celebrate his name and recognize the important role he had played in developing its multibillion-dollar business.

AFTER GOODYEAR

Vulcanization turned the sticky, smelly, sometimes hard, sometimes soft material that had cursed Charles Goodyear's life into a far more dependable material. Vulcanized rubber was stronger and lasted much longer; once treated, it did not melt and did not become hard or soft or brittle. Finally perfected in the 1850s, it arrived just in time for making tires for bicycles, which were becoming popular in the second half of the 19th century.

How Vulcanization Works

Derived from plants, rubber is an organic (carbon-based) chemical sometimes described as a natural polymer: the molecules inside rubber are very long chains made from a regularly repeating pattern of carbon and other atoms. Rubber is stretchy because these long-chain molecules are linked only loosely to one another and can be pulled apart with ease. When rubber is heated with sulfur, in the process of vulcanization, strong cross-links form between the chains.

Vulcanization makes rubber stronger because the cross-links prevent the chains from being pulled apart so easily; if the chains are stretched away from one another, the cross-links pull them back together. The greater the sulfur content, the more cross-links form and the stronger and harder the rubber becomes. Adding more than about 30 percent sulfur produces a very hard form of rubber known as ebonite (the material from which bowling balls are made). Sulfur is used to vulcanize only naturally made rubber. Artificial (chemically produced) rubbers are still vulcanized, but with different chemicals.

Although modern machines currently used for vulcanizing rubber bear little resemblance to Charles Goodyear's kitchen stove, they still work in essentially the same way. Another of Goodyear's discoveries—that adding chemicals improves the quality of rubber—also lives on. Today's rubber makers use a wide range of additives to protect rubber from gradually breaking down in sunlight and the atmosphere.

The vulcanization process adds sulfur molecules to rubber, making it able to stretch without breaking.

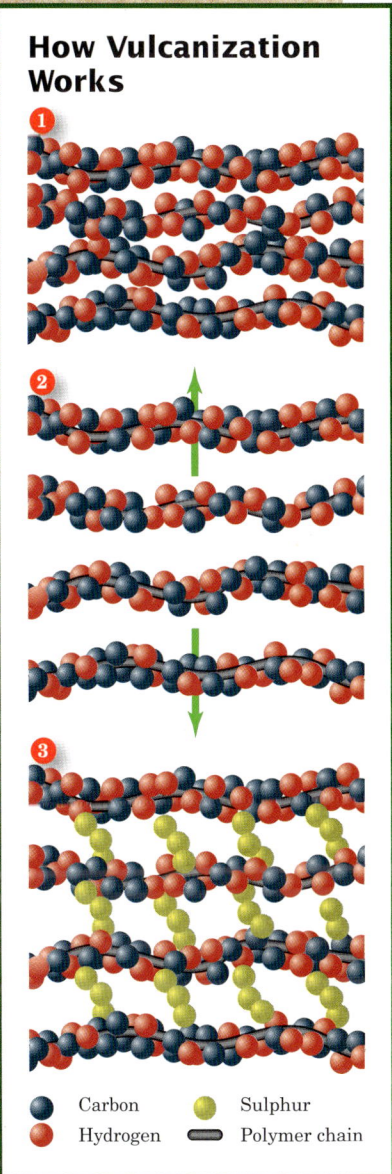

How Vulcanization Works

- Carbon
- Hydrogen
- Sulphur
- Polymer chain

When automobiles were developed at the end of the 19th century, the demand for vulcanized rubber grew enormously.

During the 20th century, chemists developed synthetic (chemically produced) rubbers that could satisfy the demand for elastic materials more cost-effectively. These modern-day industrial inventors, such as Wallace Carothers and Stephanie Kwolek, were the natural heirs of Charles Goodyear. He recognized the value of a durable, elastic material and tinkered with rubber until it could be used for many different purposes; the materials they developed started from similar inspiration. Goodyear is therefore, arguably, one of the founding fathers of modern materials science.

—Chris Woodford

Further Reading

Books
Korman, Richard. *The Goodyear Story*. New York: Encounter, 2002.

Slack, Charles. *Noble Obsession: Charles Goodyear, Thomas Hancock, and the Race to Unlock the Greatest Industrial Secret of the Nineteenth Century*. New York: Theia, 2002.

Web sites
Chemical Achievers
  Biographies of chemical pioneers from the Chemical Heritage Foundation.
  http://www.chemheritage.org/classroom/chemach/index.html

The Story of Rubber
  An online exhibition from the Polymer Science Learning Center.
  http://www.pslc.ws/macrog/exp/rubber/menu.htm

See also: Accidents and Mistakes; Buildings and Materials; Carothers, Wallace; Kwolek, Stephanie

MEREDITH GOURDINE

Inventor of device for purifying air
1929–1998

Meredith Gourdine was a physicist who specialized in the behavior of electrically charged gas particles. Despite his work in what many would consider an arcane field, Gourdine spent most of his life attempting to translate discoveries from that field into practical tools that could be used to solve everyday problems. His work resulted in a new kind of electrical generator, a new way to paint metal, and an effective method to clean the air that ultimately gave rise to the home air purifier.

EARLY YEARS

Meredith Gourdine was born in 1929 in Newark, New Jersey. He grew up in Harlem and then Brooklyn, in New York City. His mother liked math and science and his father was mechanically inclined. Like many African Americans of that era, however, Gourdine's parents were unable to obtain an education, and Gourdine's father worked as an automobile mechanic, a painter, and a janitor.

Gourdine's parents encouraged him to become educated; although he was bright, Gourdine often got bored in class and tended to be a discipline problem. When he was in the seventh grade, a teacher took Gourdine and one of his friends aside and bet them both that they would not be able to pass the math exams to get into Brooklyn Technical High School, a very competitive public school. Gourdine and his friend began studying math after class, first simply to win the bet and then because they found math interesting. They both passed the exam and were admitted. Gourdine did fairly well in math and science at Brooklyn Technical, despite holding down a job with a telegraph company at the same time. He was a good swimmer and eventually was offered a swimming scholarship to the University of Michigan.

Gourdine decided instead to attend Cornell University in Ithaca, New York, entering in 1948. It was a risky decision because Gourdine had

Gourdine was successful in a number of track-and-field events; this photo shows him coming in second in the college relay championship in April 1950. (Gourdine's uniform has a C for Cornell; the NYU runner came in first.)

no scholarship to Cornell. He had only the money he had saved from his high school job, which was enough to pay for a single semester.

CORNELL AND THE OLYMPICS

> I was always doing things people said couldn't be done.... That's the way I operated. If somebody said it couldn't be done, that's what I would try to do.
>
> —Meredith Gourdine

Gourdine wanted to study engineering physics at Cornell, but the program was extremely competitive, and his grades in high school were not good enough to gain him entrance. "That made me kind of furious," he later recalled. As a result, he worked extremely hard his first semester at Cornell, earning the grades required to transfer into the engineering physics program. His top grades also earned him a scholarship for the rest of his time at Cornell.

Gourdine also kept up with his athletics, joining Cornell's track-and-field team. He proved an exceptional athlete, winning numerous intercollegiate titles. In 1952, Gourdine competed in the Olympic Games, which were held in Helsinki, Finland. Gourdine won the silver medal in the long jump.

Gourdine completed his studies at Cornell in 1953 and joined the U.S. Navy. The navy bored Gourdine, however, and he decided to return to school. He applied for a Guggenheim Fellowship to study at the Jet Propulsion Laboratory (JPL) in Pasadena, California, which is managed by the California Institute of Technology; he won the fellowship in 1955.

ELECTROGASDYNAMICS

Gourdine received his engineering doctorate in 1960. While at JPL, Gourdine focused on the rather obscure field of electrogasdynamics. Electrogasdynamics is the study of the behavior of electrically charged gas particles, or ions, when gas is in motion.

One aspect of particular interest to Gourdine was a phenomenon that scientists had known about for centuries: moving, ionized gas can produce electricity, although too little to be of practical use. Gourdine thought he might be able to figure out a way to increase the amount of electricity this phenomenon produced, perhaps enough to create a practical electrical generator. Gourdine hit upon the idea of forcing the ionized gas through a very narrow channel. The channel lay in an electrical field, and the interaction of the field and the ions converted their kinetic energy (the energy of motion) into electricity.

The device developed by Gourdine greatly increased the amount of electricity generated by flowing gas. Gourdine was disappointed by the reaction of his supervisor at JPL, who merely complimented Gourdine on

Gourdine, third from left, with other American scientists during a meeting at Northwestern University on magnetohydrodynamics in space exploration in 1961.

his work; no attempt was made to develop the idea, and no effort was begun to find out if a practical electrical generator could be made to use the increased electricity. Although Gourdine enjoyed working at JPL, the reaction of his supervisor made him realize that an academic career was a dead end for him. The experience convinced Gourdine that he was never going to accomplish anything meaningful at the laboratory.

HIS OWN BOSS

Gourdine decided that the private sector held more promise. Although he considered starting his own company, he had a family to support; consequently, he took a job with the Plasmodyne Corporation and then with the Curtiss-Wright Corporation. Gourdine was unsuccessful in his attempts to interest his employers and other companies in his generator idea. In 1964, Gourdine borrowed money from friends and established his own company, Gourdine Systems, based in Livingston, New Jersey. Nine years later, he established a second company, Energy Innovations, in Houston, Texas.

Establishing his own firms allowed Gourdine to invent full-time. The next few decades were a period of remarkable creativity, and he was granted some forty U.S. patents between 1969 and 1996. The patents covered a wide array of applications, including reproducing images, detecting dust, cooling computer chips, and clearing fog from airport runways —all of which stemmed from his understanding of the behavior of ions.

Not all of Gourdine's ideas resulted in marketable products. Despite years of effort, including presenting the technology before a committee of the U.S. Senate in 1967, Gourdine was never able to market his electrical generator for large-scale use. Instead, he sold small models of his generators to schools and laboratories interested in studying electrogasdynamics.

Other ideas were more successful. In the late 1960s, Gourdine developed a new technology for painting metal that was widely adopted in factories (see box, Applying Paint). Gourdine's system was used for many years before it was replaced by newer technology.

CLEANER INCINERATORS

In the 1960s Gourdine became very interested in the problem of air pollution. In 1966 he served on a task force examining air pollution in New York City, which determined that the city had unsafe air, in large part because of the widespread use of incinerators that burned garbage.

Incinerator smoke contained noxious particles that were too small to be removed with a conventional filter. More advanced filtration systems were available, but they were large and very expensive. In addition, these systems typically cleaned smoke by running it through water, resulting in large quantities of wastewater.

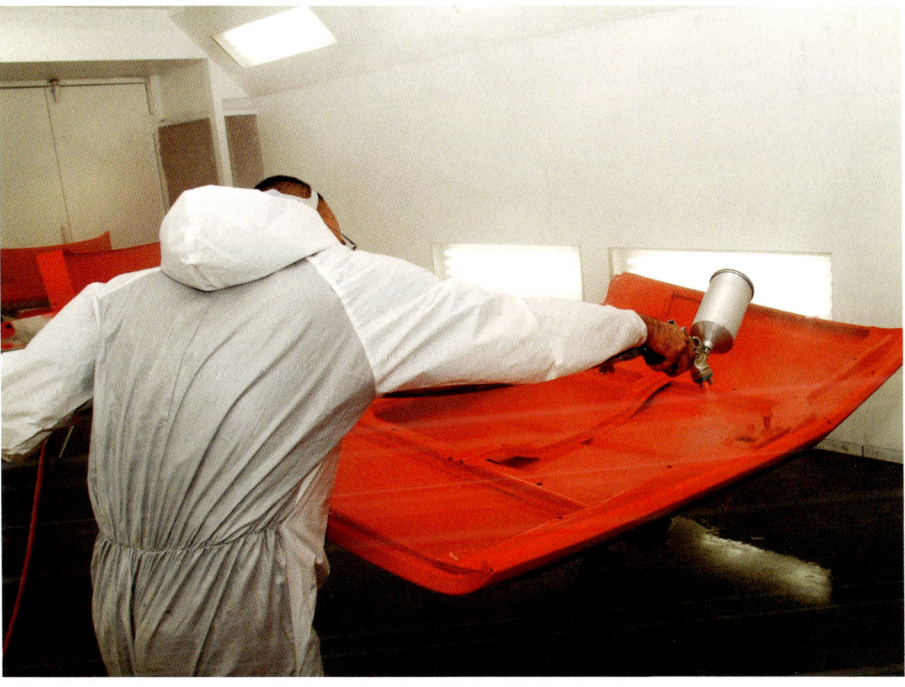

Gourdine developed a successful system for applying paint to metal, such as a car's.

Applying Paint

Gourdine's focus on the behavior of electrically charged particles resulted in the development of a new way to paint metal products such as cars, refrigerators, and other machinery. Painting such products is not simply a cosmetic issue: unpainted metals will rust when exposed to moisture. Previously, workers in an automobile factory would spray liquid paint onto a car. Spots would be missed, particularly if the body of the car had many hard-to-reach crevices. Once the car left the factory and became exposed to rain and grime, the missed spots would rust, shortening the vehicle's life.

Gourdine's system, in contrast, used dry-powder paint. The powder was given a positive electrical charge, and the automobile was given a negative electrical charge, creating a magnetic attraction between the two.

When the powdered paint was blown onto the metal, it clung to the object. Bare spots of metal had a greater attraction to excess paint powder than spots already covered with powder, so the paint powder got into every crevice. The painted object was then baked to melt the powdered paint and give it a smooth finish.

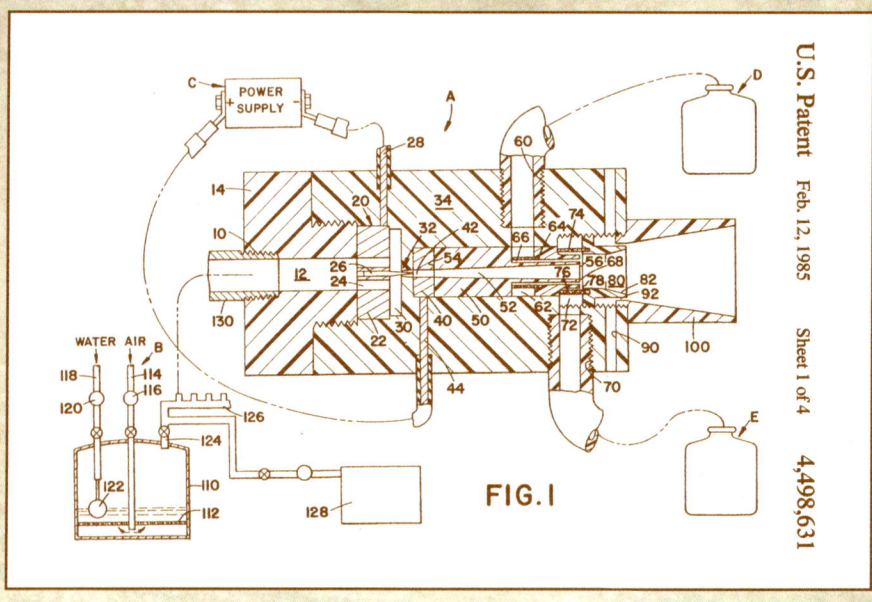

Illustrations from Gourdine's patented system for applying paint to metal.

TIME LINE

1929	1952	1960	1964
Meredith Gourdine born in Newark, New Jersey.	Gourdine wins a silver medal at the Helsinki Olympic Games.	Gourdine receives a doctorate from the California Institute of Technology.	Gourdine founds Gourdine Systems.

In the early 1970s, Gourdine developed an effective filtering device that was far smaller than existing systems. The filter worked by applying a negative charge to the particles in smoke. Once those particles were negatively charged, they passed by a positively charged metal plate. The smoke particles clung to the plate, which could then be cleaned or replaced. Gourdine's device was approved for use in New Jersey and was used on a test basis in New York City later in the 1970s. Although the filter worked well, eventually incinerators themselves fell out of favor and were banned in urban areas.

Gourdine continued working throughout the 1980s and 1990s, despite significant health problems, including diabetes, which caused him to lose his sight and a leg. He died in Houston in November 1998 at the age of 69.

Gourdine's air-filtration technology was never widely used for incinerators, but in recent years it has experienced a renaissance in the growing area of home air purifiers. Home air purifiers, which are small enough to be picked up and moved from room to room, have become popular in

Air purifiers at a Sharper Image store in New York City in 2005.

TIME LINE (continued)

1966	1973	Early-1970s	1998
Gourdine examines air pollution in New York City.	Gourdine establishes Energy Innovations.	Gourdine develops a new air filtering system.	Gourdine dies.

the United States because of concerns about indoor air pollution such as mold spores and solvent fumes released by carpets and upholstery, as well as concerns about increased rates of childhood asthma. About $430 million worth of air purifiers were sold in the United States in 2004 by major companies such as Honeywell, Sears, and Panasonic; many of these purifiers use the ionizing technology developed by Gourdine.

—Mary Sisson

Further Reading

Books

Carwell, Hattie. *Blacks in Science: Astrophysicist to Zoologist.* Hicksville, NY: Exposition, 1977.
Henderson, Susan K. *African-American Inventors II.* Mankato, MN: Capstone, 1998.
Van Sertima, Ivan, ed. *Blacks in Science: Ancient and Modern.* New Brunswick, NJ: Transaction, 1983.

Web sites

Meredith Gourdine
   A profile of the inventor from the Massachusetts Institute of Technology.
   http://web.mit.edu/invent/iow/gourdine.html
Physicists of the African Diaspora: Meredith C. Gourdine
   A biography of Gourdine from SUNY Buffalo's Mathematicians of the African Diaspora exhibit.
   http://www.math.buffalo.edu/mad/physics/gourdine_meredithc.html

See also: Buildings and Materials; Household Inventions; Science, Technology, and Mathematics.

JOHANNES GUTENBERG

Inventor of modern printing

ca. 1400–1468

Midway through the fifteenth century, a German metalworker, Johannes Gutenberg, secured his place in history when he invented the modern method of printing books. Until that time, the relatively few books in existence had been copied out laboriously by hand or were printed from hand-carved blocks, another long and difficult process. Gutenberg's invention permitted books to be mass-produced. When many more books became readily available, a greater number of people became literate and better educated; knowledge spread more quickly, bringing about huge changes in religion, the sciences, politics, and business.

EARLY YEARS

Very little is known about Gutenberg's early life. Even his most famous years, when he developed the printing press, are shrouded in mystery. It is not even known exactly when he was born, though most historians agree that it was sometime between 1394 and 1400.

We do know that Johannes Gutenberg was born into a well-off family in Mainz, then one of Germany's most important cities and a major trading center. His father, Friele Gänsfleisch zur Laden zum Gutenberg, was a wealthy merchant; his mother, Else Wyrich, was the daughter of a shopkeeper. In those times, wealthy families had names that reflected where they lived: *zur Laden zum Gutenberg* showed that Friele Gänsfleisch had two separate houses in the neighborhoods of Laden and Gutenberg. Johannes was the youngest of three children.

It is not known where Gutenberg went to school but, given his wealthy family and the knowledge he would have needed to develop his invention, he probably had a good education. Clearly he learned to read and write, since these skills were central to printing books. Some historians think he went to one of the schools linked with the 40 churches in Mainz. Others have suggested that he may have attended the nearby University of Erfurt, because many wealthy young people were educated there, including some of his cousins. At some point, Gutenberg also learned metalworking, either from his father or from an uncle, but he

A page from the Diamond Sutra, made in 868.

would not have served an apprenticeship, because he was an aristocrat. Although many jewelers and goldsmiths were based in Mainz at that time, the Gutenbergs played a special role: one branch of the family had long been responsible for the mint (making coins by stamping designs into metal). This knowledge may have inspired Gutenberg's method of printing.

Gutenberg lived at the end of the Middle Ages, the period of history that started around 400 CE. Few books were available and most people were illiterate (it is estimated that only one person in 20 could read or write in Germany at this time). Most of the books were large religious volumes such as Bibles, each of which had to be copied by hand from an earlier book. These books were called manuscripts (literally, "handwritten"); the "scribes" who produced them were extremely skillful, but not necessarily that well educated (some could not themselves read or write). Some of the books they produced, known as illuminated manuscripts, were elaborately decorated with colorful inks and opulent designs in gold and silver. Each book took months or years to make and was very valuable, not least because it was unique.

Printed books did exist in the Middle Ages, but mainly in Asia. The first printed book, known as the *Diamond Sutra*, was made in China in 868 CE. Each of the book's seven pages was carved from a separate flat block of wood, which was then covered with ink and pressed firmly onto paper so the carved design appeared in reverse. Once the block was made, it could be used to print a single page of a book over and over again. Sometimes very elaborate books were made with block printing. In the year 972, Chinese printers made the Tripitaka, a Buddhist sacred text. Each of its 130,000 pages was carved by hand from a separate wood block. Because carving small letters into wood accurately was difficult, block printing was most suitable for printing designs on picture cards, which were very popular at this time when so few people could read words. However, this type of block printing never became popular in Europe.

The big drawback of block printing was that many new blocks had to be prepared for each new book—a slow, time-consuming, and expensive process. The solution was to use not one large block to print a page but many smaller blocks to print each of the letters on that page. These blocks are called pieces of type; many different pages can be printed using the same type simply by rearranging the same letters to make different words. This invention, movable type, first appeared in China around the year 1000 and in Korea around 1300, but it was not a practical invention at that time. Asian languages use many thousands of

> In the late 1430s, Gutenberg sold small mirrors (or "looking glasses") to the many people who attended a major exhibition of religious relics. Before people looked around the exhibition, they would pin a mirror to their hats in the hope that the mirror would catch some of the relics' magical properties and could transfer these to their friends and relatives at home.

picture symbols instead of our 26 letters, so a Chinese or Korean printer would still have needed thousands of tiny wooden blocks to print a book in this way.

Asia had developed a remarkable printing method but could not really use it efficiently; Europe had a growing need for books but no real way to produce them quickly. When Johannes Gutenberg "reinvented" movable type in Europe in the 1400s, he satisfied this need and caused a revolution. It is unlikely, though not impossible, that Gutenberg knew movable type had already been invented in Asia. Like many other inventions, it was an idea that was rediscovered and reinvented through years of painstaking experiments.

THE GUTENBERG BIBLE

By 1430, Gutenberg had moved to the city of Strasbourg, where he seems to have tried a variety of jobs, including working as a police officer and a teacher (he taught a colleague, Andreas Dritzehn, to cut and polish gemstones). Around this time, Gutenberg also began to experiment with better ways of printing.

Although records are sketchy, historians do not think Gutenberg's invention was a eureka moment—a brilliant stroke of genius that simply came to him in an instant. Like many modern inventors, he seems to have perfected his ideas by trial and error over many years, using several existing inventions together to make a completely new printing process. He began by printing from single, large, solid pieces of wood

A statue honoring Gutenberg stands in his home town of Mainz, Germany.

The block printing technique is still used in some instances, such as in these New Year pictures made at the Tongshunde art shop in Weifang, China, in 2006.

(block printing) and went on to experiment with printing using individual letters made from tiny pieces of metal (type printing). He developed a way of moving those metal pieces of type to print many different pages, a technique called movable type. He also devised a new type of ink and a way to use a wine-making screw press to print pages evenly and reliably. With these innovations, modern printing was born (see box, How Gutenberg's Press Worked).

By 1450, Gutenberg had returned to Mainz and was ready to start work on his first major book. This came to be known as the Gutenberg Bible, Mazarin Bible, or 42-line Bible (because each page contained exactly 42 lines of words). It was a very ambitious project: historians believe around 180 copies were produced, each containing 1,282 pages bound into two separate volumes. Gutenberg clearly wanted to produce a book that was at least as impressive as a medieval Bible, so each copy was also illuminated (illustrated) and rubricated (marked with red headlines) by hand. The mammoth project seems to have been completed

How Gutenberg's Press Worked

"A spring of truth shall flow from [my printing press]. . . ."

Imagine Gutenberg's workshop—the heat of molten metal, the smell of wet ink, the rustle of thick paper, the regular squeak of the printing press. This humble place in the heart of Mainz was the birthplace of a new way of printing that changed human civilization perhaps more than any other invention in history. Gutenberg himself wrote, "A spring of truth shall flow from [my printing press] . . . . Like a new star it shall scatter the darkness of ignorance, and cause a light heretofore unknown to shine amongst men."

Gutenberg did not invent printing, but in the course of 20 years of experimenting, he managed to improve every single stage of the printing process. He printed each page of his famous Bible using thousands of small pieces of metal type, one for each character (letter, number, or punctuation mark) on the page. Using small wooden wedges (quoins), he arranged the pieces of type snugly in a wooden frame (form), brushed the form with ink, and then pushed it firmly against a sheet of paper using a press. By covering the form with more ink and repeating the process, Gutenberg could make an exact copy of the same page. By rearranging the type to make new words and sentences, he could print different pages.

Before Gutenberg could print anything, he had to make his type. To make one letter, he used a sharp tool to engrave (carve) the shape of that letter into a piece of hard metal. This was called a punch, because Gutenberg used it to punch the letter shape into a flat piece of copper (a softer metal) called a matrix. Next, he placed the matrix into the bottom of a mold and filled it with hot molten (liquid) metal. When this cooled, it turned solid and formed (cast) a piece of type. He used the same mold to cast many identical versions of the same letter; he used similar molds to cast all the other letters he needed. The words Gutenberg printed in this way looked more like handwriting than a modern book, because he was trying to make printed books that resembled old, handwritten Bibles.

The press Gutenberg used was a modified winepress. The ink-covered form, containing the type, rested on a wooden block; the paper was on top of the form; and another wooden block (called a platen) was moved on top of the paper. Gutenberg used an enormous screw on the top to tighten up this sandwich of materials and press the paper onto the inked type. The advantage of using a press was that it gave equal pressure across the page and produced an even print.

Gutenberg also made his own very thick printing ink. Carbon gave the ink its black color, and small amounts of metals such as copper and lead gave it a shiny quality. The ink was oil-based, not water based, so it would stick to the metal type more easily. On some copies of his Bibles, Gutenberg also printed the red headlines called rubrics, but using two colors of ink proved slow and difficult, so this job was later done by hand.

Gutenberg did not, however, make his own paper: he used handmade paper imported from Italy or vellum (a high-quality, long-lasting material made from calfskin; it is estimated that the skin of 170 calves would have been needed for each copy of the Bible). Gutenberg printed his famous Bible in a style called folio. He printed sheets of paper with two pages on each side and folded them down the middle to make one leaf. He collected around ten of these leaves to make a gathering, and he stitched about thirty gatherings together to make two volumes, each of about three hundred pages.

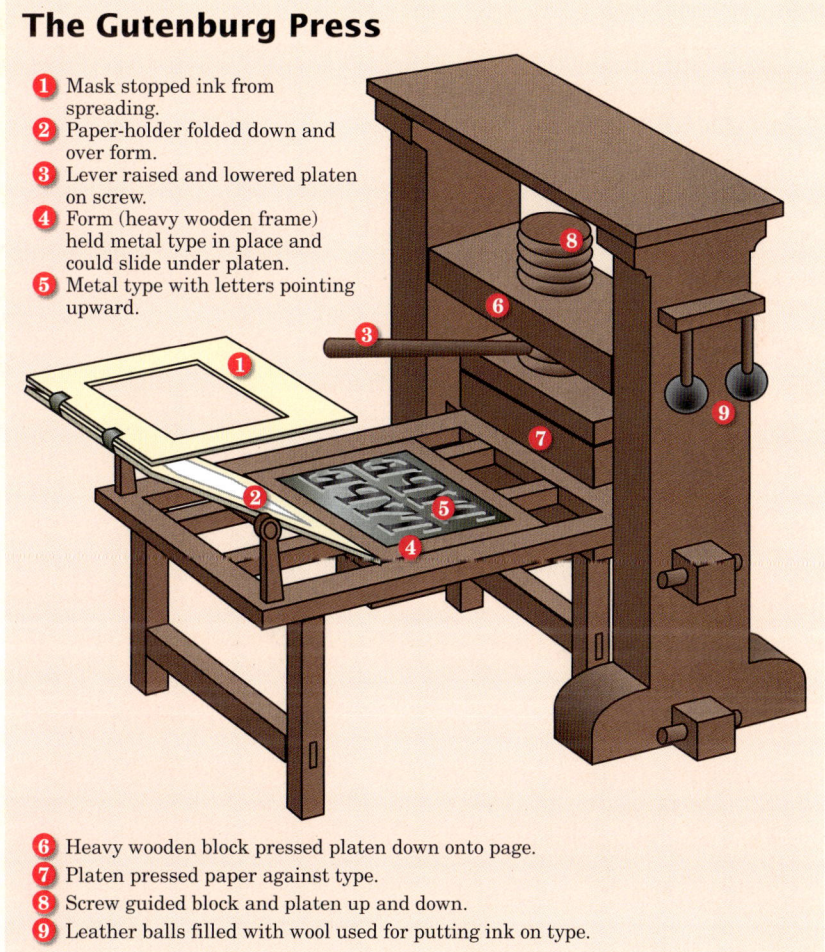

The Gutenburg Press

1. Mask stopped ink from spreading.
2. Paper-holder folded down and over form.
3. Lever raised and lowered platen on screw.
4. Form (heavy wooden frame) held metal type in place and could slide under platen.
5. Metal type with letters pointing upward.
6. Heavy wooden block pressed platen down onto page.
7. Platen pressed paper against type.
8. Screw guided block and platen up and down.
9. Leather balls filled with wool used for putting ink on type.

The parts of Gutenberg's printing press.

between 1453 and 1456; Gutenberg also worked on other publications at the same time, including a Turkish calendar that he printed in 1454.

DOWNFALL

Like many other inventors, Gutenberg had to raise huge sums of money to finance his project. Constructing his printing equipment, particularly the many thousand letters of type that had to be cast out of metal, was very expensive. He managed to raise the needed funds by going into business with a wealthy moneylender, Johann Fust (ca. 1400–ca. 1466). Gutenberg had promised that they would see great profits from the business, but the three loans he took out from Fust, totaling around two thousand gulden (coins), were substantial—much more than an average worker at that time could earn in his entire life.

Gutenberg's downfall happened for a reason that is all too familiar in the history of invention. Fust was eager to see the profit from his investment, but Gutenberg was constantly trying to improve his idea and needed ever more money to do so. When Fust finally lost patience in 1455, he sued and forced Gutenberg to surrender his part of the firm. Without Gutenberg, Fust and his associate Peter Shäffer went on to publish a beautiful Psalter (book of psalms) in 1457, with two-color printing and two different sizes of type. Many people believe Gutenberg was responsible for these innovations.

Although Gutenberg was ruined, he managed to set up another printing firm and continued working in the Mainz area for several years. Historians disagree about his later life. Some say that he lived in poverty, but it seems he may have been more fortunate. In January 1465, he found favor with Adolph II, the archbishop of Mainz, who granted him a pension including a yearly allowance of grain, cloth, and 2,180 liters (2,304 qt) of wine for his own use. He died in February 1468, about age 70, and was buried at the Franziskus church, Mainz.

An undated etching of Johannes Gutenberg.

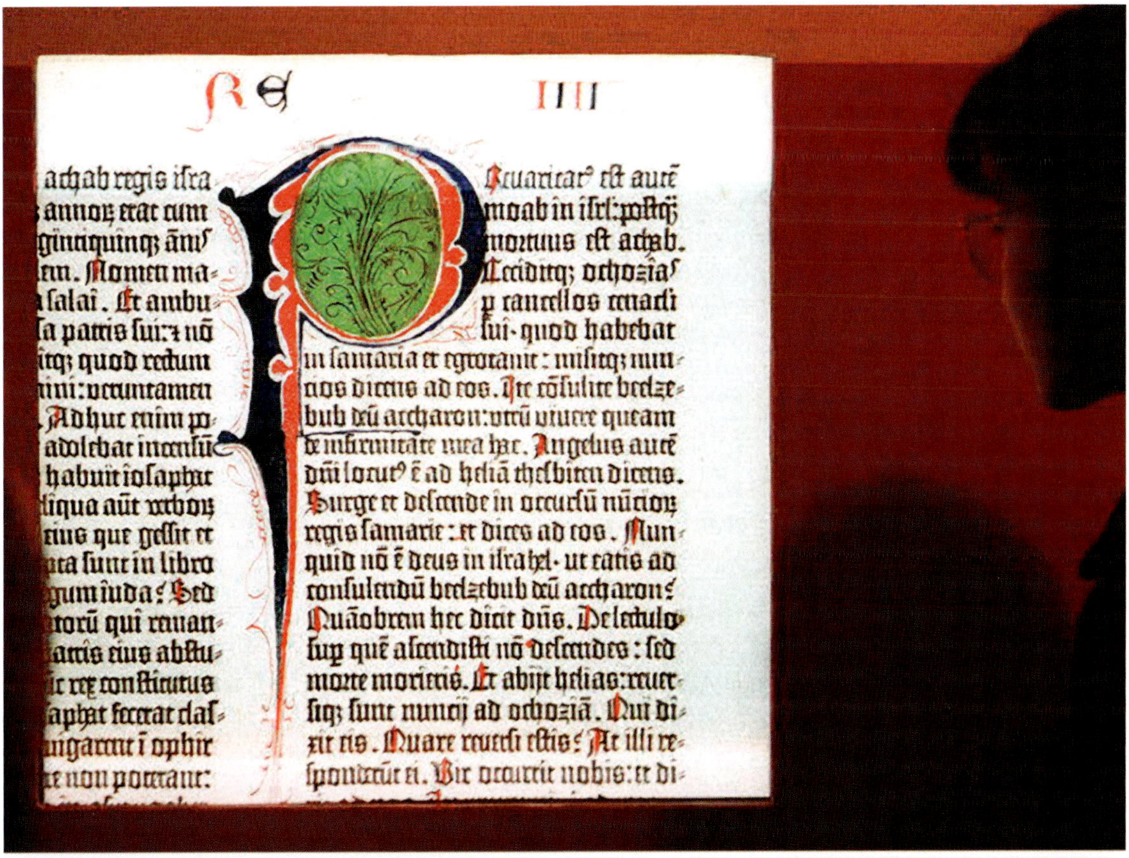

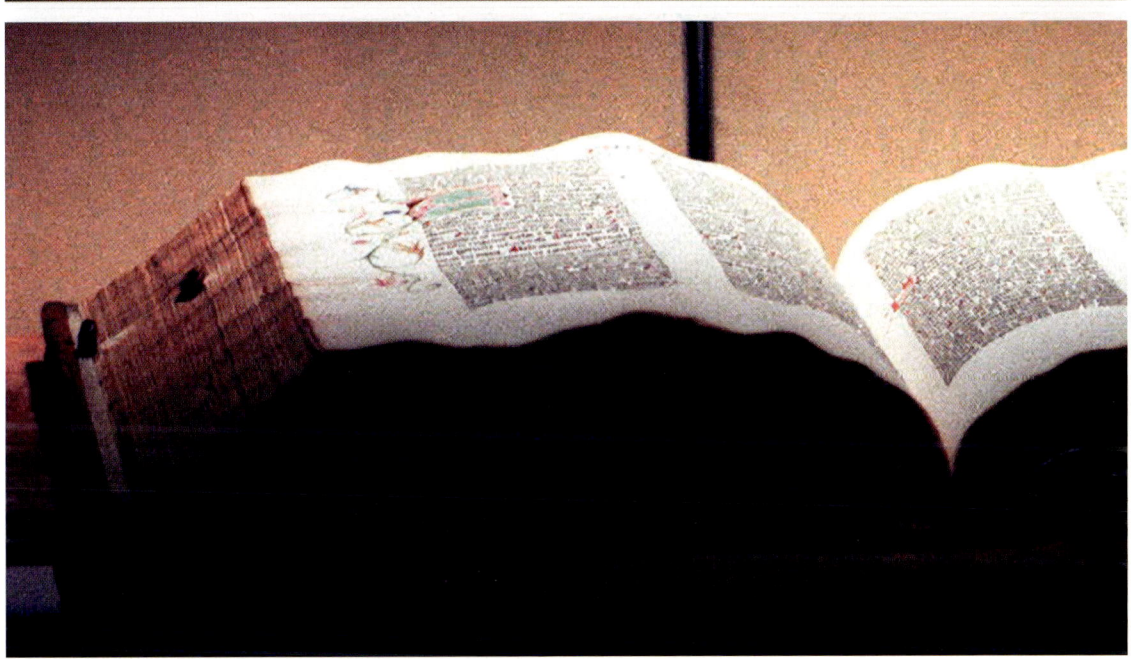

An original Gutenberg Bible (bottom) and the same Bible re-created through digital technology (top) at an exhibition in Tokyo in 2003.

THE IMPACT OF THE PRINTING PRESS

The 50 years after Gutenberg developed his new printing method saw a huge increase in the number of printing presses in operation and books published. Books ceased to be one-of-a-kind works of art; suddenly, they were mass-produced items. Historians think that more than six thousand different titles were published during this time, including around forty thousand separate copies of the Bible. The technology spread rapidly through Europe, largely thanks to German printers who took their printing skills to other nations. The first American printing press was established in Mexico City in 1539.

Gutenberg's development of movable type was the most important technological advance in the Renaissance (literally, "rebirth"), the period of history roughly between 1400 and 1600 when many new advances in art, religion, and science were made. Books spread knowledge and allowed people to share ideas widely for the first time. Most of these early books were still published by the Roman Catholic church, but people began to examine and discuss religious ideas much more closely and critically.

Mechanical presses allow books and newspapers to be produced on a scale unthinkable in Gutenberg's time.

Among those scrutinizing the Catholic church and its theology were Martin Luther (1483–1546) and others; their criticisms of certain practices of the church and their alternative readings of sacred texts led to a huge period of religious upheaval known as the Reformation during which Protestant churches were established. Books helped to spread scientific ideas and knowledge, too, often pitting new scientific discoveries against the religious views of the day. By the 1600s, printing was also being used to produce newspapers and pamphlets, with an enormous effect on the spread of trade and political ideas around the world.

TIME LINE

ca. 1400	1430	1450	1453–1456	1465	1468
Johannes Gutenberg born in Mainz, Germany.	Gutenberg moves to Strasbourg and begins to experiment with printing.	Gutenberg returns to Mainz.	Gutenberg produces the Gutenberg Bible.	Gutenberg receives a yearly pension from the archbishop of Mainz.	Gutenberg dies.

Printing advanced even further during the 19th-century Industrial Revolution, when the arrival of sturdy steam-driven presses made from iron allowed books and newspapers to be copied faster and in larger quantities than ever before. Hand-operated presses, such as Gutenberg's, could make no more than a few hundred copies per hour; by the 1870s, a roller press developed by American printer Richard March Hoe (1812–1886) was printing up to 18,000 newspapers in the same time. Another great advance was the linotype machine, developed in 1884 by a German-born American printer, Ottmar Mergenthaler (1854–1899). It allowed printers to make up (or "set") a whole line of type automatically from molten metal, greatly accelerating the printing process. Printing using movable metal type continued until the 1970s and 1980s, more than 500 years after Gutenberg, when electronic and photographic forms of copying and publishing gradually replaced it.

Gutenberg did not invent printing, books, or movable type. His genius was to develop a new way of printing books quickly, reliably, and in large quantities. That technology allowed ideas and knowledge to take root and spread as never before, and the steady spread of knowledge prompted significant changes in society. In many ways Gutenberg's invention was quite a humble one, but its effect on human civilization could hardly have been more dramatic.

—Chris Woodford

Further Reading

Books

Burch, Joann Johansen. *Fine Print: A Story about Johann Gutenberg.* Minneapolis, MN: Carolrhoda, 1992.

Fisher, Leonard Everett. *Gutenberg.* New York: Atheneum, 1993.

Fussel, Stephan. *Gutenberg and the Impact of Printing.* Translated by Douglas Martin. Cornwall, UK: MPG, 2005.

Man, John. *Gutenberg: How One Man Remade the World with Words.* Hoboken, NJ: Wiley, 2002.

Web sites

British Library Digital Collection: Gutenberg
    Information about Gutenberg's Bible and online images of pages from the book.
    http://www.bl.uk/treasures/gutenberg/homepage.html

Gutenberg Museum
    A museum dedicated to Gutenberg in the German town of Mainz.
    http://www.gutenberg-museum.de/

International Printing Museum
    Information about printing methods and printing presses through history.
    http://www.printmuseum.org/

See also: Communications.

CHARLES HALL

Developer of aluminum
manufacturing process
1863–1914

Because of its lightness, malleability, and resistance to corrosion, aluminum has become one of the most widely used metals on earth. Until the 1880s, however, aluminum was used only for luxury goods like jewelry and tableware. That changed when Charles Hall, a young man working in a small town in Ohio, developed a method to cheaply convert the aluminum oxides that naturally occur in soil into pure, useful aluminum metal.

EARLY YEARS

Charles Martin Hall was born in Thompson, Ohio, on December 6, 1863. His father was a minister, and for much of Hall's childhood, his family moved to various towns in Ohio as Hall's father moved from church to church.

When Hall was 10 years old, his family settled permanently in Oberlin, Ohio. Hall's father had graduated from Oberlin College, and he expected Hall and his six other children to attend it. The college was a coeducational institution, which was a rare at that time. Sending so many children to college on a minister's salary posed a serious challenge to the Halls—money was always tight, and as a teenager Hall worked at odd jobs to help support his family.

Hall was fascinated by chemistry at an early age, reading and rereading one of his father's chemistry textbooks. As a teenager, he became interested in becoming an inventor in the field of science. Perhaps because of his family's poverty, Hall was also interested in the potential

Undated photograph of Charles Hall.

financial benefits of inventing, and he familiarized himself with the workings of patent law. In letters to his siblings, in particular to his older sister Julia, who shared his enthusiasm for chemistry, Hall would note ideas for inventions. He then instructed the recipient to be sure to save the letter so that he, Charles, would have proof that the idea was his.

ALUMINUM

One of the most intriguing—and potentially lucrative—scientific puzzles of the day was the problem of purifying aluminum. Aluminum, an element that is found abundantly in the soil (about 8 percent of the earth's crust is aluminum), never occurs in nature in its pure form.

Aluminum was first isolated by scientists in 1825, but not until 1845 was enough aluminum purified to allow the metal to be studied. Research indicated that pure aluminum had some extremely useful properties: it was exceptionally light, and it was malleable enough to be easily worked when cold. Further research showed that aluminum was extremely resistant to corrosion. The element's highly reactive nature is the reason: when pure aluminum is exposed to air, it immediately reacts with oxygen, forming a tough, thin layer of oxide around the metal that prevents any further corrosion.

Throughout the latter half of the 19th century, scientists continued to develop new, cheaper methods of purifying aluminum. The price of aluminum dropped from more than $500 per pound in 1852 (more than twice the cost of gold or platinum) to about $12 per pound in 1862 (about the same as silver). By the 1880s, aluminum cost about $8 per pound—cheaper than silver. However, even with the new methods, aluminum was still far too expensive for any large-scale use. Aluminum was used in jewelry, and in 1884 a cap for the Washington Monument in Washington, D.C., was made from aluminum, but it had few practical applications.

THE WOODSHED

Hall began attending Oberlin College in 1880 at the age of 16. There, he met chemistry professor Frank Jewett, who was also interested in the aluminum question. Jewett had a sample of the metal in its pure form—which was still rare—that he would show students.

With Jewett's encouragement and assistance, Hall began serious efforts to purify aluminum. He took a year off from college in 1883, during which time historians believe he concentrated on his aluminum experiments. By his senior year of college, Hall was focusing his attention on using electricity to purify aluminum (see box, How Hall Purified Aluminum).

How Hall Purified Aluminum

Aluminum in nature is mostly found as oxides, which form when electrons are taken away from aluminum atoms. Hall's experiments centered on restoring electrons to those atoms—a chemical process that is called reduction.

An electrical current is produced by forcing electrons to move from one point to another, a process that frees the electrons to react chemically. Hall put two electrodes into water containing aluminum oxides and ran a current through it. His hope was that the positively charged aluminum oxides dissolved in the liquid would be attracted to negative electrodes and that the oxides then would pick up the electrons needed to form pure aluminum.

Hall was dissolving his aluminum oxides in water, and the water was reacting with the mix, so that pure aluminum was never formed. Hall decided that he needed to dissolve the aluminum oxides in a liquid that was not water and that would not attract water from the air. He began to focus on fluoride salts, which required melting at an extremely high temperature. He needed several tries before he found a fluoride salt that he could melt. In early February 1886, Hall finally managed to melt the fluoride salt cryolite.

A few days later, Hall melted cryolite in a clay-lined tube, mixed in the aluminum oxide, and ran an electrical current through the resulting liquid. He let the mixture cool and then broke the tube apart. The only result was a gray deposit on the negative electrode that was clearly not aluminum. Hall decided that he needed to line the tube with something that would not react chemically with the mix. Hall settled on a graphite liner, and he modified the cryolite so that it would melt at a lower temperature. On February 23, 1886, he melted the cryolite and aluminum oxide in the graphite-lined tube and passed an electrical current through it for several hours. Hall then let the mixture cool and solidify, broke it apart with a hammer, and found several small, silver-colored globules of pure aluminum.

By the time Hall graduated from Oberlin College in 1885, he still had not managed to purify aluminum. He continued his efforts in a woodshed attached to the back of his family's house, working in fairly primitive conditions. Hall manufactured many of his own chemicals—to get the electricity he needed for his experiments, he had to make his own batteries.

At the time very little was known about the basic properties of the chemicals that Hall was working with; this lack of information forced him to conduct a lot of his own basic research. With his sister Julia's assistance, on February 23, 1886, Hall finally discovered the correct mix of chemicals, heat, and electricity needed to manufacture pure aluminum.

THE PATENT PROBLEM

Hall had made an important breakthrough, but he had nowhere near enough money to fund his efforts to turn his aluminum-purification process into a large-scale industry. With the help of his family, Hall immediately began the search for investors, even traveling to Boston to get backing. He also hired a patent attorney, who applied for a patent for Hall's purification process in July 1886.

A few months later, Hall received notice from the U.S. Patent Office that someone else had already applied for a patent for the same process. A French scientist, Paul Héroult, had applied for a patent in France in April 1886 and for a U.S. patent the next month. Both men had independently discovered the same method for purifying aluminum; as a result, the method is often called the Hall-Héroult process.

Fortunately for Hall, he and Julia had kept detailed records of his work, and Hall had also written letters to family members informing them of his progress. As a result, Hall was able to prove to the U.S. Patent Office that he had made his discovery in February 1886, before Héroult had filed, and he received the patent in April 1889.

Illustration from Hall's patent for his aluminum purification process.

FINDING A BACKER

In the meantime, Hall still needed funding to develop and expand his method of purifying aluminum. For a time, he worked for the Cowles Electric Smelting and Aluminum Company, which had developed a method to make aluminum alloys without making pure aluminum first. Hall felt, however, that his research was not supported at Cowles and became convinced that the company was trying to prevent him from commercializing a potentially competing technology.

Hall left Cowles in July 1888, but while he was with the company he had met Romaine Cole, a man who had previously worked for Alfred Hunt, a well-connected Pittsburgh metallurgist and businessman. Cole told Hunt about Hall's process, and Hunt became very enthusiastic about the project. By the end of the summer, Hunt had gathered together investors and chartered a company, the Pittsburgh Reduction Company, to commercialize Hall's process.

Hall came to Pittsburgh and began perfecting his process. Within six months Pittsburgh Reduction was producing about fifty pounds (23 kg) of aluminum daily. At the time, Pittsburgh was a burgeoning steel town, and Hall had little trouble there getting support for the idea of mass-producing aluminum. In early 1890, Pittsburgh Reduction won the financial backing of the wealthy Mellon banking family, which gave it the resources it needed to expand.

Aluminum alloy being prepared for casting at a plant at Aluminum Industries, Inc., in Cincinnati, Ohio, in 1942.

MASS-PRODUCING ALUMINUM

Many manufacturers that potentially could use aluminum had years of experience working with other metals, and they were reluctant to take the time and trouble of learning how to manufacture goods with aluminum. As a result, Pittsburgh Reduction was quickly forced to expand its operations from creating raw aluminum to manufacturing more finished aluminum goods, including foil, sheets, rods, wire, and cable.

One of the first major consumer uses for aluminum was kitchen pots and pans. The aluminum pots and pans stayed shiny and bright, but they were often poorly made. Pittsburgh Reduction felt that the shoddy products were hurting aluminum's reputation; thus the company eventually began manufacturing its own pots and pans under the brand name WearEver.

Aluminum's light weight did win it some early adherents. Among them were two bicycle mechanics, Orville and Wilbur Wright (1871–1948 and 1867–1912). In December 1903, the Wright brothers flew the first airplane, which was outfitted with a motor made of aluminum. To this day, airplanes are typically made of aluminum.

ALCOA

Hall remained with Pittsburgh Reduction for the rest of his life, continually improving his process. In 1907, the company, which had expanded to several locations across the United States, changed its name to Aluminum Company of America. By 1910 employees were calling the

Aluminum is a common component of airplanes and cars. General Motors workers inspect the all-aluminum GM 3.6L V-6 VVT engine in 2006.

TIME LINE

1863	1885	1886	1889	1890	1907	1914
Charles Martin Hall born in Thompson, Ohio.	Hall graduates from Oberlin College.	Hall discovers how to purify aluminum.	Hall receives patent for his aluminum-purification process.	Pittsburgh Reduction wins the financial backing of the Mellon family.	Pittsburgh Reduction changes its name to Aluminum Company of America (Alcoa).	Hall dies.

company "Alcoa" for short, a name the company increasingly used for itself until it officially changed its name to Alcoa in the 1990s.

In 1908, Hall, who had moved to Niagara Falls, New York, to be near a new manufacturing plant, began feeling weak and unwell—the beginnings of a fatal bout with what historians believe was probably leukemia. Hall battled ill health for several years, spending the winters in warmer climates. The Society of Chemical Industry awarded Hall the Perkin Medal, considered America's highest honor for industrial chemistry, in 1910.

In 1914, Hall wintered in Daytona, Florida, where he died that December, leaving an estate of at least $30 million. The majority of his fortune was left to Oberlin College. Alcoa remains one of the world's top producers of aluminum, achieving $30 billion in sales in 2006. Aluminum itself is one of the most widely used metals in modern times and can be found in vehicles, electrical wiring, homes, and airplanes.

—Mary Sisson

Further Reading

Books

Edward, Junius. *The Immortal Woodshed: The Story of the Inventor Who Brought Aluminum to America.* New York: Dodd, Mead, 1955.

Smith, George David. *From Monopoly to Competition: The Transformations of Alcoa, 1888–1986.* Cambridge: Cambridge University Press, 1988.

Web site

Production of Aluminum Metal by Electrochemistry
An exhibit on Hall's work by the American Chemical Society.
http://acswebcontent.acs.org/landmarks/landmarks/cmh/index.html

See also: Buildings and Materials; Wright, Orville, and Wilbur Wright.

LLOYD A. HALL

Inventor of food preservation methods

1894–1971

Lloyd A. Hall was a prominent chemist who devoted his life to developing superior methods for curing and preserving various foods. From the 1920s until the end of his life, Hall invented techniques that remain important to the food industry today. In the process, he received more than 100 patents in the United States and elsewhere.

EARLY YEARS

Lloyd Augustus Hall was born in Elgin, Illinois, on June 20, 1894. In the 1830s his paternal grandfather had been a founding member, and later pastor, of the Quinn Chapel A.M.E. Church—the first African American church in Chicago. His maternal grandmother had come to the city at about the same time via the Underground Railroad, the system of paths and homes that helped escaped slaves from the South find refuge in the North. Hall's father, Augustus Hall, also became a minister in a Chicago Baptist church. Both his father and his mother, Isabel, had graduated from high school in the Chicago area.

The family lived in nearby Aurora, where Lloyd Hall attended East Side High School. Active in athletics and captain of the debate team, he graduated near the top of his class and was offered scholarships to several universities in Illinois. He selected Northwestern University in Evanston, just north of Chicago. As a chemistry student there, he befriended a fellow student, Carroll L. Griffith, whose family would later found Griffith Laboratories. That association would later play a very important role in Hall's career. After graduating from Northwestern in 1916, he began taking graduate courses in chemistry at the University of Chicago.

Hall took a job in the laboratories of the Chicago Department of Health in 1917 and quickly became a senior chemist. He also served briefly during World War I inspecting explosives in an ordnance department in Wisconsin. Hall worked in various industrial concerns for a number of years before he joined the Boyer Chemical Laboratory in Chicago as a chief chemist in 1921. At Boyer, Hall first developed his interest in the emerging field of food chemistry. One year later, he took a job as president of a consulting lab in Chicago. He worked as an independent consultant for the next several years.

Undated portrait of Lloyd A. Hall.

GRIFFITH LABORATORIES

In 1924 Carroll Griffith, Hall's lab partner from Northwestern, offered Hall the use of his laboratories while Hall carried out his consulting work. What began as an informal arrangement soon turned into the most lasting and valuable association in Hall's career. In 1925 Hall became director of research at Griffith as well as the company's chief chemist. He continued independent consulting until 1929, when he devoted himself full-time to Griffith Laboratories. He remained with the company until his retirement in 1959.

Griffith Laboratories was one of the pioneers in bringing scientific innovation to the food industry. Lloyd Hall, already interested in

The Flash-Drying Method

Soon after joining Griffith, Hall sought a better way to preserve and cure meats. The most common way of preserving up to this time (the mid-1920s) was to use sodium chloride—common table salt. Curing meats involves using chemicals, often nitrogen compounds such as nitrates or nitrites, in combination with salt to further preserve the color and appearance of salted meat. By coming up with a method of flash drying these salt and nitrogen compounds, Hall solved an old problem: how to allow the sodium chloride to penetrate the meat first to preserve it adequately, before the nitrate compounds would enter. All previous experience demonstrated that nitrates and nitrites penetrated meat faster than salt, causing the meat to disintegrate before it had been properly preserved.

Hall speculated that if nitrate compounds could be combined with salt to form a third substance, then they could penetrate the meat at a rate that would allow proper preservation and maintain the original color, appearance, and texture. He sought to achieve this combination by very quickly drying (flash drying) a solution of nitrate compounds and sodium chloride over hot metal rollers. The crystals that resulted from this evaporation consisted of a nitrate center surrounded by a salt crust. This was just the structure needed: as the crystals dissolved, the salt would penetrate the meat first. Hall's crystals quickly became the most popular meat-curing product on the market.

improving the foods available to consumers, found an ideal environment at Griffith for carrying out his research. Although he was involved in many of Griffith's important projects, Hall is remembered chiefly for three major improvements for which he is responsible: creating a type of sodium crystal for preserving meats; sterilizing spices; and preventing rancidity, or spoiling, in fats and oils.

ADVANCES IN PRESERVING MEATS

Hall's first major contribution to food preservation came with improving the existing method of preserving and curing meats. Using a "flash-drying" approach, he developed a kind of crystal compound, made of common table salt and sodium nitrate, that allowed the salt to penetrate the meat, thus preserving it, while the nitrates acted to further cure the meat, maintaining its original color and texture (see box, The Flash-Drying Method).

Hall's advances in meat preservation quickly caught on in the food industry. With Hall at the helm of research, Griffith Laboratories gained a wider reputation as a leader in food chemistry. Other major advances soon followed.

STERILIZING SPICES

In addition to meat-curing crystals, Hall created a process for sterilizing spices, thus ridding them of any contaminants. All kinds of dried spices, as well as dried vegetables such as onion powder, were subject to attack by various bacteria and yeasts and their shelf life was, accordingly, fairly short. No satisfactory means of ridding these spices of impurities without ruining their taste and appearance had yet been found.

Hall experimented with various approaches, including heat-drying spices in the air and baking them in an oven, but none proved effective

TIME LINE

1894	1916	1921	1925	1959	1962	1971
Lloyd Hall born in Elgin, Illinois.	Hall graduates with chemistry degree from Northwestern University.	Hall joins Boyer Chemical Laboratory in Chicago.	Hall becomes director of research at Griffith Laboratories.	Hall retires from Griffith.	Hall works with the American Food for Peace Council.	Hall dies.

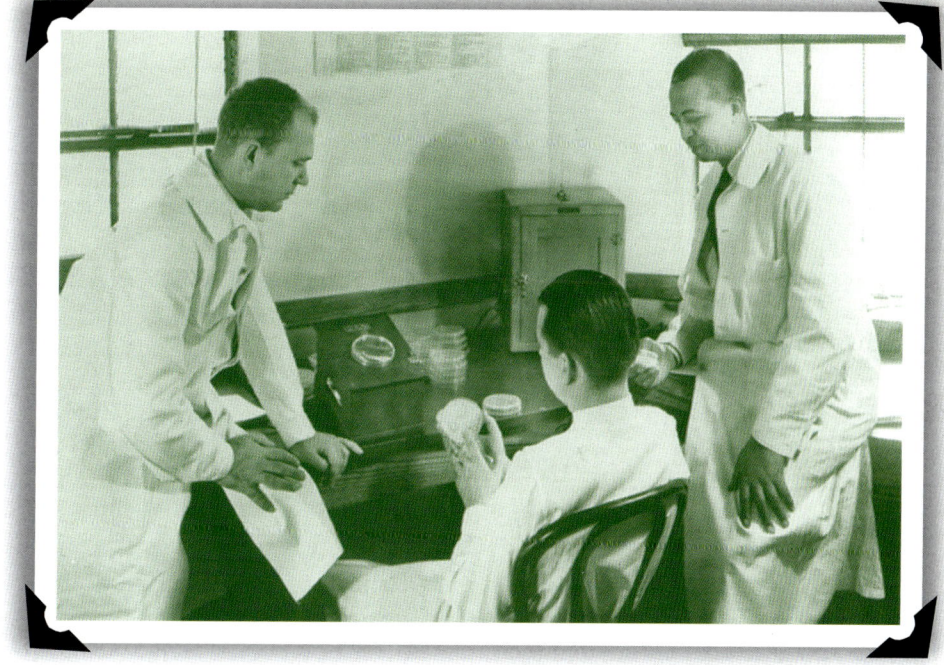

Lloyd Hall (far right) and colleagues at Griffith Laboratories (undated photograph).

while also maintaining flavor and color. He finally arrived at a solution by using a kind of gas, ethylene oxide, to kill the contaminants and germs in spices. He placed spices in a vacuum chamber to remove their moisture and allow the gas to enter the spices. The method Hall developed to sterilize spices is still used to sterilize medical instruments and some medicines.

PREVENTING SPOILAGE

A third major advance Hall made in food chemistry involved the use of antioxidants to prevent fats or oils from becoming rancid, or spoiling. Fats or oils in food can become rancid when they come into contact with oxygen.

Hall found that antioxidants occurring naturally in crude, unrefined vegetable oil could be mixed with salt to create a substance that would prevent fats and oils from spoiling. The antioxidants kept these fats from reacting with oxygen, thereby giving them a longer shelf life.

OTHER ACCOMPLISHMENTS

Hall played an important role in many other advances within the food industry. Notable among these was his work to reduce the time for curing bacon from a couple of weeks to only a few hours, in the process improving the bacon's quality and taste. He won a patent for this invention—one among many over his career. He served on many local, state,

Workers trim beef briskets, which will be cured in brine and turned into corned beef, at the Vienna Beef factory in 2006, in Chicago, Illinois.

and national boards, including the Chicago chapters of the National Association for the Advancement of Colored People (NAACP) and the Urban League (both devoted to equality for African Americans); the Illinois State Food Commission; the Science Advisory Board on Food Research; and the Institute of Food Technologists, of which he was also a cofounder.

After he retired from Griffith Laboratories in 1959, Hall served several months in Indonesia as consultant for the United Nations' Food and Agricultural Organization unit. President John F. Kennedy appointed him in 1962 to the American Food for Peace Council, where he remained until 1964. Throughout the rest of his life, Hall served in a variety of organizations and causes in the same spirit that had led him to invent improved food preservation methods more than 30 years earlier. In his final years, Hall worked to help underprivileged young people get the training they needed to work in chemistry. Hall died on January 2, 1971, in Altadena, California, where he had moved with his wife following his retirement. Hall was inducted posthumously into the National Inventors Hall of Fame in 2004.

—Paul Schellinger

Further Reading

Book
Haber, Louis. *Black Pioneers of Science and Invention*. New York: Harcourt Brace Jovanovich, 1970.

Web site
Black Inventor Online Museum: Lloyd Hall
 A biography of Hall hosted by Adscape International.
 http://www.blackinventor.com/pages/lloydhall.html

See also: Birdseye, Clarence; Borlaug, Norman; Food and Agriculture.

JAMES HARGREAVES

Inventor of the spinning jenny
ca. 1720–1788

James Hargreaves was an illiterate weaver who worked in an obscure textile industry in a remote part of England. Nonetheless, he developed a truly revolutionary machine—the spinning jenny—which set into motion a series of events that would change first his industry, then his country, and then the world.

EARLY YEARS

Little is known of Hargreaves's early life, in part because he never learned to read or write and so left no firsthand accounts. Hargreaves spent much of his life in Lancashire County in northwest England, a fairly isolated part of the country where illiteracy was common and record keeping was poor. Baptismal records indicate that he was most likely born in 1720 near the town of Blackburn. There is no evidence that he ever attended school. Twenty years after his baptism, Hargreaves married, and ultimately he was father to roughly a dozen children.

Like many in the Lancashire area, Hargreaves was a weaver of cotton. At the time, England's economy was largely agricultural, and historians speculate that Hargreaves probably farmed or raised livestock in addition to weaving.

WOOL VERSUS COTTON

In the early 18th century, wool was the dominant British textile, as it had been for centuries, but Lancashire textile workers specialized in making cotton fabric. Unlike wool, which came from sheep that flourished in England, cotton would not grow in England and had to be imported. Lancashire's unusual focus on cotton may have developed because of the county's proximity to ports that handled much of the country's international shipping. In addition to raw cotton, which was imported from the eastern Mediterranean, the Caribbean, and the United States, high-quality cotton cloth was imported from India.

Spinning wool into thread.

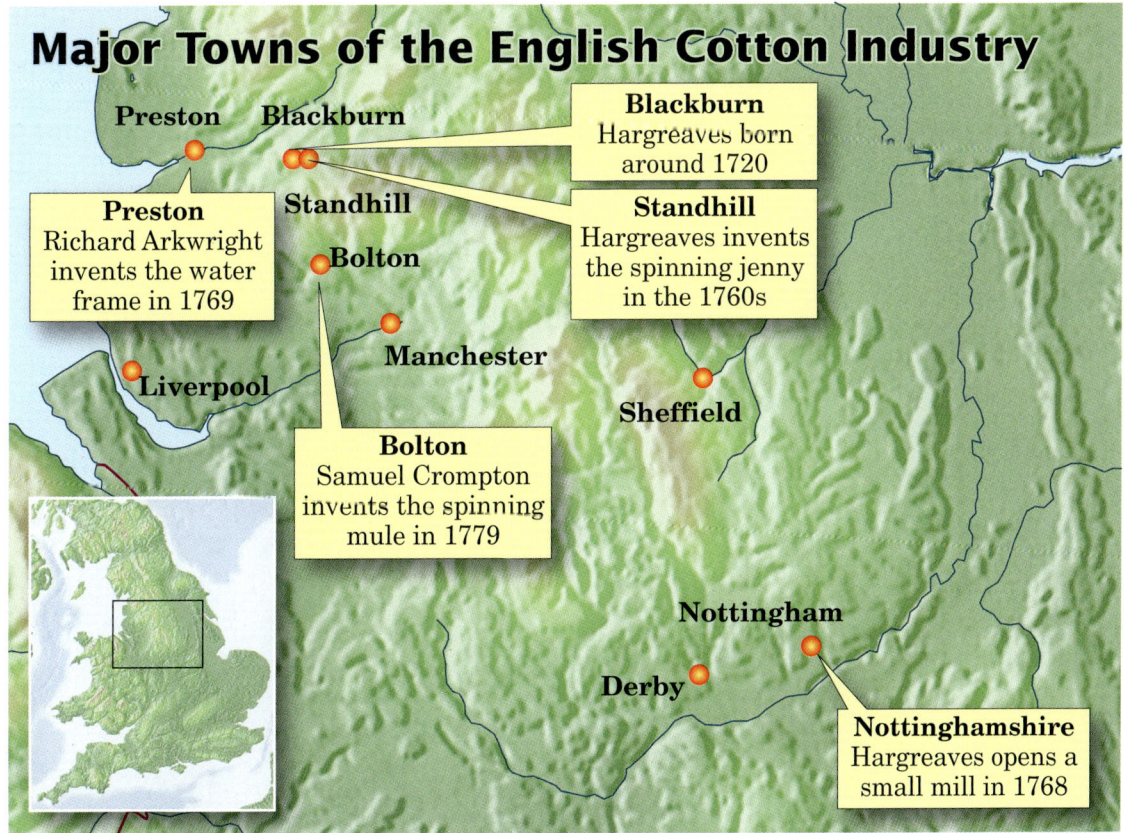

These cotton-producing towns were at the heart of the cotton industry in 18th- and 19th-century England. Most of them are located in Lancashire County, in the northwest of England.

Because cotton came from abroad, it was often regarded as a threat to the British woolen industry, which supported countless British shepherds as well as textile workers. Indeed, a writer in the 1630s quoted a popular saying about the town of Pilton, where cotton fabric was woven: "Woe unto you Piltonians that make cloth without wool."

The hostility toward cotton was such that, in 1721, the British government banned the purchase, sale, and wearing of printed cotton goods. In 1736 the government eased the ban slightly, allowing the manufacture of fustian, a type of cloth that was woven from both cotton and linen thread. Because of such restrictive policies, cotton, although important to Lancashire, made up only a tiny portion of the total British textile industry.

IMPROVING TECHNOLOGY

Ironically, cotton's foreign origin would contribute to its ultimate dominance of the British textile industry. Unlike weavers of wool, cotton workers did not have centuries of tradition dictating how to do their jobs. Although resistance to change was certainly present, the cotton industry as a whole was much more accepting of new technologies than were other textile industries.

Despite his lack of formal training in carpentry and machinery, Hargreaves had a strong interest in those disciplines, and at some point—records do not indicate exactly when—he began to invent new tools to help cotton workers do their jobs. Hargreaves developed an improved method of carding (preparing raw cotton to be spun into thread). In traditional carding, workers would lay raw cotton onto a wire brush, then comb it by hand with another brush, creating coils of cotton (roving) that would have all the fibers running in the same direction.

The small handheld wire brushes made carding slow work. Hargreaves conceived a setup that allowed workers to use much larger brushes. One brush, which held the raw cotton, was placed on the floor, while a second brush was suspended from the ceiling. A system of pulleys allowed a work-

Spinning Thread

Spinning is the process by which fibers are turned into usable yarn or thread. A substance like cotton in its natural state is made up of a mass of fibers that are too short to be very useful in the manufacture of cloth, so the fibers must be enmeshed to be made longer. A spinner twists and pulls roving (cotton that has been combed so that all its fibers are aligned with one another) to produce thread. If this is done incorrectly, the fibers do not enmesh together well, and the thread breaks easily. Spinners can use a variety of methods to create thread. In the fourteenth century the spinning wheel became enormously popular in Europe and ultimately became the basis for Hargreaves's spinning jenny.

A spinning wheel has a large wheel connected to a smaller projection, called a spindle, by a belt. Because the wheel is big and the spindle is small, a few revolutions of the wheel (which is usually powered by a foot pedal, or treadle) will make the spindle revolve many times.

The spindle usually has two devices attached that revolve with it: a U-shaped device (flyer) and a spool. The spinner hooks one end of the roving to the spool, the flyer, and the tip of the spindle, and then begins to spin the wheel, turning the spindle and its spool and flyer.

The flyer twists the roving, while the spinner holds and pulls it. Once the roving has been twisted and pulled into a length of thread, the spinner lets the thread wind onto the spool and goes to work pulling and twisting the next section of roving.

> From the year 1770 to 1788 a complete change had gradually been effected in the spinning of yarns. That of wool had disappeared altogether, and that of linen was also nearly gone; cotton, cotton, cotton, [became] the almost universal material for employment.
>
> —William Radcliffe, 1828

er to comb through the cotton using the suspended brush. Because the brushes Hargreaves invented were so much larger than earlier brushes, a worker could card cotton at twice the previous rate.

In the 1760s, Hargreaves turned his attention to speeding the process of spinning (see box, Spinning Thread). In the previous decade, the cotton industry had adopted a new weaving technology that had been developed for woolens: the flying shuttle. Like woolen cloth, cotton cloth was woven by laying out a set of threads, called the warp, onto a loom. Using a a shuttle, a weaver would interlace a second set of threads, called the weft, at a right angle to the first. Before the invention of the flying shuttle, a weaver had to insert the shuttle through the warp by hand. This limited the width of the fabric, which could be only as wide as the weaver—or a weaver and an assistant—could comfortably reach. The flying shuttle was a device that used hammers to knock the shuttle through the warp, making possible the weaving of wider fabrics without an assistant.

The flying shuttle made weaving much faster. Ironically, this worsened an existing problem: spinners could not keep up with the demand for thread from weavers. Even before the invention of the flying shuttle, the work of three or four spinners was needed to make enough thread to keep one weaver busy. The problem was so severe that by the late 1730s a machine had been invented that attempted to spin thread by passing roving through rollers; it did not work well, however, and was never widely adopted. As the flying shuttle became more popular in the 1760s, the shortage of thread became increasingly critical.

THE SPINNING JENNY

The spinning jenny, invented by Hargreaves, helped spinners to produce thread in greater volume. Although some historians claim that the spinning jenny was invented in 1767, most think it was invented a few years before, in 1764.

The jenny operated much like a traditional spinning wheel; thus spinners trained on the old wheel could usually convert easily to the new machine. It had one large power wheel operated by a treadle. Instead of having a vertical wheel with one spindle positioned horizontally, however, the jenny had a horizontal wheel with eight spindles positioned vertically. A single spinner using such a jenny could produce eight times as much thread as a spinner using a traditional wheel.

A reconstruction of Hargreaves's original spinning jenny.

At first Hargreaves kept his invention secret, building jennies only for family members and friends. Eventually, however, word of the new invention spread. At the time England was in an economic depression, and other spinners in the area began to worry that this new machine would force them out of work. A mob gathered, broke into Hargreaves's home, destroyed all the jennies they could find, and caused extensive damage. This violent incident forced Hargreaves to move to Nottinghamshire in 1768; there, he and a partner opened a small jenny mill.

Hargreaves had not patented his invention—at the time, a patent cost at least £100, a sum few could afford. Hargreaves's jenny was soon widely copied, so that obtaining a patent became impossible. In 1780 Hargreaves was finally able to patent a 16-spindle jenny; nonetheless, he never received much in the way of financial reward for his invention. Hargreaves had only a modest fortune when he died eight years later in Nottinghamshire. By that time, some twenty thousand spinning jennies are estimated to have been in use.

WATER FRAMES AND SPINNING MULES

The success of the spinning jenny led others to develop spinning machines. In 1769 Richard Arkwright (1732–1792) patented a water frame, a device that relied on the older technology of rollers to turn roving into thread.

Arkwright most likely did not actually invent the water frame, and although he took out several patents for new kinds of textile equipment, all of them were invalidated in 1785. Nonetheless, he perfected the water frame and made the use of rollers practical. Unlike the jenny, the water frame was too big to be powered by a person. Instead, the frame had to be powered by water (hence the name) and thus could be operated only inside a mill. Arkwright opened the first mill that used water frames in 1771 near Nottinghamshire. When that proved successful, he

A Cottage Industry

In the 1700s, weavers like Hargreaves did not usually work in factories or mills. Indeed, factories in general were quite rare at that time. Instead, the textile industry—an industry that at the time employed more people in England than any other occupation except agriculture—was largely a cottage or domestic industry.

Instead of coming to a mill to work each day, textile workers typically worked at home. The machinery in use at the time was small and relatively inexpensive, so it was fairly easy for a person even of modest means to buy the necessary equipment and set up a shop either outside or in a spare room.

Textile workers usually received their raw material from and returned the finished product to a merchant, who would then pay them for their work. This arrangement had several variations: sometimes the workers would rent their equipment from the merchant as well, and sometimes the workers would purchase raw material and sell the finished work on their own. The merchant would then sell the finished cloth to retailers.

Manufacturing finished cloth from raw fiber involves many steps. The fiber must be carded, or combed, into roving; the roving must be spun into thread or yarn; the thread must be woven into rough cloth; and then the rough cloth must be dyed and finished. Typically, different individuals undertook each step in the process, although they might all live under the same roof. For example, the looms used to weave cloth were heavy and hard to operate, so weavers were usually men in good physical condition—like Hargreaves. Spinning was easier, so children, women, and the elderly usually spun.

opened mills all across Great Britain, becoming remarkably wealthy—and powerful enough to get the ban enacted on all-cotton fabric in 1721 lifted in 1774.

While cotton mills began cropping up in England, the home-based (or cottage) industry did not suddenly vanish (see box, A Cottage Industry). Instead, for decades the two types of cotton manufacturing coexisted—the water frame in mills and the spinning jenny in homes. Indeed, spinning jennies became larger and larger: in 1784, 80-spindle jennies were in use, and by 1800, the number of spindles on a single spinning jenny had increased to 120.

Despite their widespread use, both the jenny and the water frame had significant limitations. The thread produced by the spinning jenny was not very strong and could be used only as weft in weaving. Thread made with a water frame was stronger, but coarse. In 1779 a young man, Samuel Crompton, who had worked on a spinning jenny as a teenager, hit upon the idea of combining the two technologies. Crompton's invention—a cross between a water frame and a spinning jenny—was called a spinning mule for its combination of techniques (a mule is the

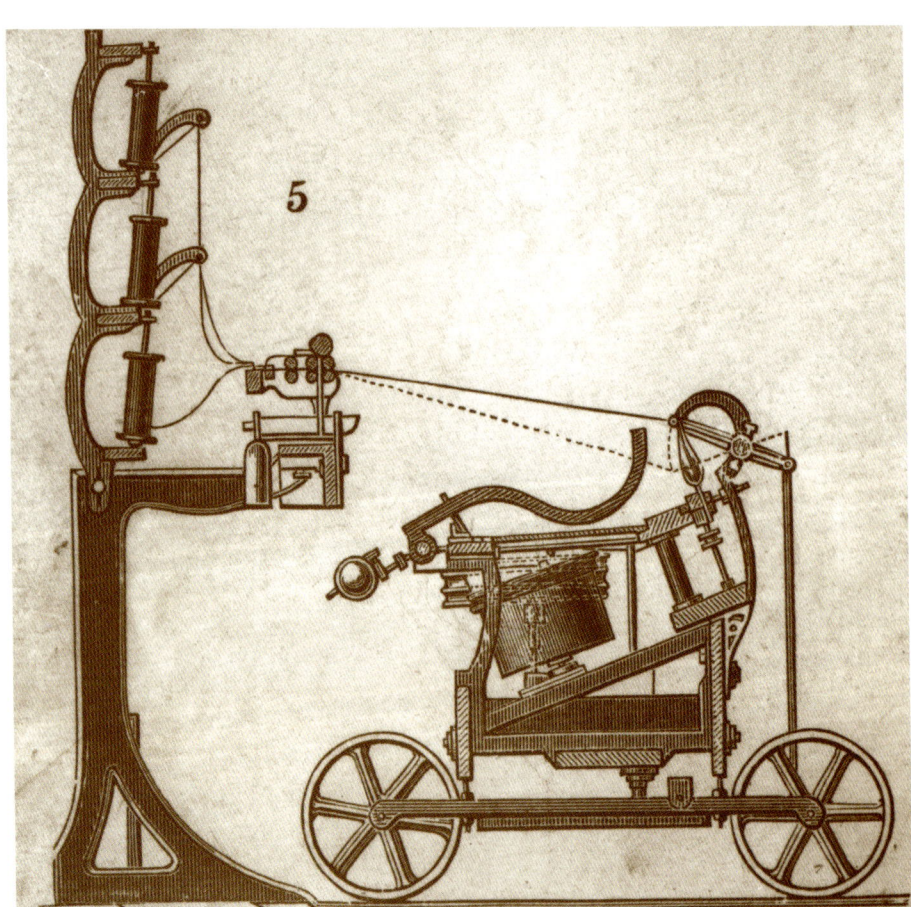

Undated illustration of a spinning mule, invented by Samuel Crompton in 1779.

TIME LINE

ca. 1720	1740	ca. 1764	1768	1780	1788
James Hargreaves born near Blackburn, England.	Hargreaves marries.	Hargreaves invents spinning jenny.	Hargreaves moves to Nottinghamshire and opens a mill.	Hargreaves patents his 16-spindle jenny.	Hargreaves dies.

offspring of a horse and a donkey). In a spinning mule, cotton roving was first fed through rollers, as in a water frame, and then was spun on spindles, as in a jenny. The result was a strong cotton thread that could be spun coarse or fine, depending on what was desired.

Spinning mules were flexible in another way—they could be made large enough to use in a mill, where they were powered by water and eventually steam, or small enough to be used at home. By the 1810s the mule had largely replaced both the water frame and the jenny. In modern times spinning mules themselves have largely been replaced by faster automatic spinning machines, but the mules are still used in Great Britain and elsewhere to manufacture yarn for high-end knitwear.

THE INDUSTRIAL REVOLUTION

The development of the spinning jenny and spinning mule, combined with other advances in carding and weaving, and the lifting of the ban on all-cotton cloth, led to an expansion of the cotton industry in Great Britain. The British imported only 5 million pounds (2.2 million kg) of raw cotton in 1781, but nine years later, the amount had increased to 31 million pounds (14 million kg). In 1820, Great Britain imported 150 million pounds (68 million kg) of raw cotton.

A weaving factory in Dumfermline, Scotland, around 1880.

Far from destroying the British economy, cotton was providing the country with lucrative new products—cotton cloth and cotton thread—that could be sold abroad. The value of Great Britain's cotton exports, which had been a mere £355,000 in 1780, shot up to £5.4 million in 1800 and £20.5 million in 1820. By

1830, cotton exports accounted for fully half of the value of all British exports, and more efficient methods of manufacturing the product had led to a sharp drop in the price of cotton.

The explosive and unexpected growth of the cotton industry was widely noted in Great Britain. As people in other industries sought to expand, they looked to the cotton industry and followed its example. The use of increasingly specialized workers, of a centralized factory or mill, and of machines like the spinning jenny that allowed fewer people to do more work ultimately altered the economy first of Great Britain, then Europe, and then the United States, from one based mostly on agriculture to one based on manufacture—a shift that became known as the Industrial Revolution. Although the transition was not without complications, ultimately the Industrial Revolution would contribute to a remarkable improvement in living standards in western Europe and the United States.

—Mary Sisson

Further Reading

Books
Bythell, Duncan. *The Handloom Weavers: A Study in the English Cotton Industry during the Industrial Revolution*. Cambridge: Cambridge University Press, 1969.

Deane, Phyllis. *The First Industrial Revolution*. Cambridge: Cambridge University Press, 1965.

Gregg, Pauline. *A Social and Economic History of Britain 1760–1970*, 6th ed. London: George G. Harrap, 1971.

Web sites
Cotton Town
   A historical exhibit on the British cotton industry by the town of Blackburn with Darwen.
   http://www.cottontown.org

Spinning the Web: The Story of the Cotton Industry
   A collection of digitized items from libraries and archives in northwest England, developed by the Manchester Library and Information Service.
   http://www.spinningtheweb.org.uk

See also: Cloth and Apparel; Jacquard, Joseph-Marie; Whitney, Eli.

JOHN HARRISON

Inventor of the chronometer
1693–1776

To avoid hitting rocks, to steer the shortest course between two points, or simply to avoid getting lost, ships at sea must know their exact position, both by latitude (how far north or south they are) and by longitude (how far east or west). Seafarers and others had been able to figure out their latitude for centuries, but no one had a practical way of finding longitude until John Harrison solved the problem by telling the time.

EARLY YEARS

John Harrison was born on March 24, 1693, in Foulby, Yorkshire, in northern England. After his family moved about forty miles (65 km) east to Barrow, Lincolnshire, he trained as a carpenter and taught himself how to make clocks by repairing old ones. Between the ages of 19 and 23 he made three pendulum clocks, which still survive. As a carpenter, he made most of the internal parts from wood instead of the usual steel and brass, which were more expensive and would have had to be made by a metalsmith.

With a growing local reputation, Harrison was invited in 1720 to build a tower clock for the stables of Brocklesby Park in Lincolnshire. In this clock, which still runs in its original location, Harrison eliminated steel altogether. It also runs without oil or grease, because many of the internal parts are made of lignum vitae, a tropical American hardwood that has enough natural oil to need no further lubrication. In the mid-1720s Harrison crafted two grandfather clocks with further innovations, which he said made them so accurate that they lost or gained no more than a second per month. The best clockmakers of the time in London and Paris boasted of making clocks that were accurate only to 10 seconds per month. Now Harrison was ready to tackle the problem of a clock that would work at sea and could thus be used to determine longitude.

TACKLING THE LONGITUDE

In contemporary times, a ship can get its position with a handheld receiver that gathers signals from Global Positioning System (GPS) satellites. In the 18th century, navigation was much trickier, requiring difficult astronomical observations and complicated calculations. Longitude was particularly hard to determine (see box, Latitude and Longitude), and many people thought an easier way could be devised to figure longitude. Finding a way to determine longitude was considered so urgent that in

An oil painting of John Harrison with his marine chronometer, created by Thomas King in 1767.

1714 the British government offered a reward of £20,000 (equivalent to about $5 million today) to anyone who could come up with a practical solution (plenty of impractical ones already existed).

A pendulum clock would not work, because the ship's rolling motion would interfere with the swinging of the pendulum. Pocket watches driven by springs rather than pendulums and weights had been available since the early 1500s, but two centuries later even the best were still accurate only to within a few minutes a day—much less accurate than pendulum clocks and nowhere near accurate enough to determine longitude. A timepiece that could be used to tell

Latitude and Longitude

Before the invention of GPS, to find the latitude of a position (how far north or south of the equator it is) one had to look at the sky. Around noon each day, a sextant (a handheld device that measured the angle from the horizon to an object in the sky) could be used to determine how high the sun was. At noon, the sun would be at its highest point in the sky. This information could be used to set a watch to the local time; in addition, by knowing both the date and how many degrees above the horizon the sun was, one could also determine one's position north or south of the equator. However, sailors also needed to know their longitude (how far east or west they were). Only when both latitude and longitude coordinates are known would individuals be able to locate their position on a map.

Longitude, however, cannot be determined just by observing the sun. Those needing to determine their longitude had to observe when the earth's moon passed in front of stars along its path or when the moons of Jupiter passed behind that planet. The timings of these events at a particular place—for example, Paris—were calculated in advance, and then printed in tables. By comparing the local time at which they observed these events with the times in the table, sailors could calculate how far east or west of Paris they were by the time difference; each hour of difference in times was equivalent to 15 degrees of longitude. However, observing faint, tiny stars or moons at night was almost impossible. If a clock could tell what the time was in Paris, a sailor could calculate longitude without any need for nighttime observations.

the longitude had to lose or gain no more than a few seconds over several weeks.

Harrison's solution was to replace the swinging pendulum with levers, balanced on a pivot in the middle, like a vertical seesaw. Instead of letting gravity do the work, a pair of these levers were mounted side by side and attached to each other by a spring on each end. The tension in the springs pulled first on one end, then the other, so that the levers swung back and forth in mirror image with perfect regularity, regardless of the angle. Harrison took five years to build this clock, known today as H1, completing it in 1735.

A trial voyage from Britain to Portugal and back showed that the clock kept time to within a few seconds a day—far better than all but the finest land clocks of the time. However, Harrison was unhappy with it and wanted to make a smaller version. By the time he finished this smaller version, which he called H2, in 1741, he could already think of further improvements, and he did not even bother to ask for a sea trial. Receiving regular grants from the government to continue his work, Harrison spent the next 18 years in virtual isolation creating H3, in which the balance bars were replaced by balance wheels, further insulating the clock from the effects of lurching and tilting while at sea.

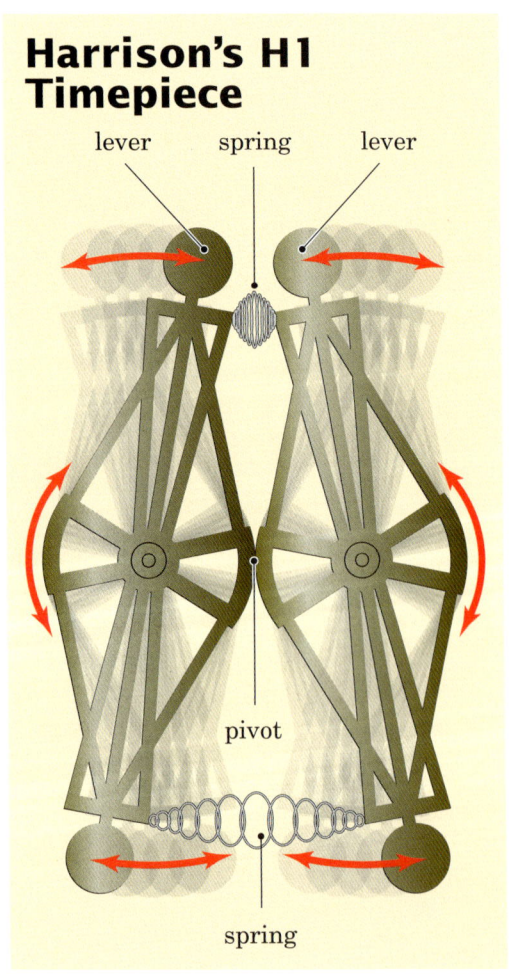

Harrison's H1 Timepiece

THE WATCH

However, at some point during those long years Harrison must have started thinking and working in a new direction; in 1760, only one year after unveiling H3, he produced his fourth timepiece, which was not a refinement of the previous three clocks but something completely different: H4 is a watch. Only 5 inches (12.5 cm) across and weighing 3 pounds (1.36 kg),

Harrison's H1 timepiece replaced the pendulum with a mechanism that made the clock independent of gravity.

John Harrison's H4 marine chronometer.

as compared with H3's 60 pounds (27 kg), H4 used vastly different mechanisms. The watch was powered by a coiled spring that was wound on a new kind of escapement (the mechanism that regulated the rate at which the spring wound down), and featured tiny, perfectly cut diamonds and rubies instead of caged ball bearings to reduce friction. Somehow, Harrison had mastered all these new technologies on his first attempt and achieved an accuracy matching that of the best clocks. Because of its accuracy and practical size, H4 is generally considered to be the most important single timepiece ever made. It is also the world's first chronometer—a handheld timepiece accurate enough to use for calculating coordinates on a long voyage.

Harrison was finally satisfied enough with his work to allow H4 to be sent on a full trial at sea from Britain to the Caribbean and back. In the 81 days the ship took to make the outward voyage, H4 lost just five seconds. Harrison was required to make two copies of H4 to prove that it could be reproduced and that its accuracy was not just a fluke, but by the time he completed the first of these, H5, he was 79 years old and told the government that he could not begin work on another one. The British government still did not award the prize, however, partly because of the influence of Nevil Maskelyne, the astronomer royal; he supervised the trials of Harrison's timepieces, but Maskelyne was himself an advocate of the alternative lunar-distance method. Harrison petitioned King George III, who himself participated in the testing of H5 and who ensured that in 1773 Parliament awarded Harrison a sum equal to the prize money. Harrison died on March 24, 1776, his vision of finding the longitude by means of a timepiece having been fulfilled.

In 1772, British naval explorer James Cook set off on the second of his three famous voyages to the Pacific with a copy of H4 on board. When he returned three years later, Cook said that the watch had "exceeded the expectations of its most zealous advocate." In the decades that followed, accurate maps began to be made of many of the world's most inaccessible

TIME LINE

1693	1714	1720	1735	1741	1760	1776
John Harrison born in Foulby, England.	The British government offers a large reward to whoever can solve the longitude problem.	Harrison is invited to build a tower clock for the stables of Brocklesby Park.	Harrison builds the H1 timepiece.	Harrison completes the H2 timepiece.	Harrison unveils his greatest achievement, the H4 timepiece, the world's first chronometer.	Harrison dies.

coastlines, mountain ranges, and polar wildernesses, all because travelers could now fix their locations precisely anywhere on earth. Harrison's chronometer, which made these maps possible, forever changed navigation, exploration, trade, and transportation.

—Jonathan Dore

Further Reading

Books

Landes, David S. *Revolution in Time: Clocks and the Making of the Modern World.* Cambridge, MA: Belknap, 1983. Rev. and enlarged edition, Cambridge, MA: Harvard University Press, 2000.

Sobel, Dava. *Longitude: The True Story of a Lone Genius Who Solved the Greatest Scientific Problem of His Time.* New York: Walker, 1995.

Web sites

Latitude and Longitude
   NASA page explaining the ideas behind coordinates and how they form the basis of timekeeping and time zones.
   http://www-istp.gsfc.nasa.gov/stargaze/slatlong.htm

Worshipful Company of Clockmakers
   Resource page on the company and its Guildhall museum in London, which holds Harrison's early clocks and H5.
   http://www.clockmakers.org/page2.html

See also: Getting, Ivan I.; Transportation.

HEALTH AND MEDICINE

Inventions have helped to revolutionize medicine. Scientific instruments such as thermometers, stethoscopes, microscopes, and x-rays have given physicians a way to gain a clearer picture of how the human body works. Improvements in surgery, the invention of vaccination, and the development of pharmaceuticals have saved millions of lives. The latest advances in genetics are expected to bring dramatic new benefits that could tackle illnesses such as cancer and make people healthier than ever before.

MEDICINE BEFORE SCIENCE

Scientific and technical knowledge have increased systematically since ancient times, but medical knowledge has often lagged behind. In all parts of the world, early attempts to cure illnesses were based on folklore and religious beliefs; and natural plants such as herbs, selected by trial and error, were the first medicines (see box, Inventing Drugs).

The ancient Greeks, who developed some of the earliest scientific ideas, also made some of the first real attempts to determine the underlying causes of illness and disease. Sometimes known as the father of medicine, Greek philosopher Hippocrates (ca. 460–377 BCE) refused to accept that diseases were punishments sent by the gods, as people had once thought. Thinking like a scientist, he wrote: "Every disease has its own nature, and arises from external causes." Galen (ca. 130–200 CE) was another pioneering doctor of ancient times. By dissecting the bodies of apes, he helped to found the sciences of anatomy, which looks at the major structures inside the body; and physiology, which considers the more detailed inner workings of these structures. The

Undated portrait of Greek physician Hippocrates.

HEALTH AND MEDICINE

Inventing Drugs

Drugs are among the oldest medical inventions. In India, ancient physicians used hundreds of different medicines made from animal substances such as milk and bone, as well as from vegetables and minerals. The ancient Chinese developed a similar system based on herbal remedies that is still widely used. Some ancient remedies are still used in western medicine. One of the world's most popular drugs, aspirin, is made from a chemical found in the bark of the willow tree, which was known to ancient Greeks and Native Americans as a painkiller. It was first made as a drug in 1897 by German chemist Felix Hoffmann (1868–1946) to help relieve his father's arthritis.

The antibiotic penicillin, discovered by Alexander Fleming in 1928, proved to be a revolutionary drug treatment for many illnesses caused by germs. Another modern drug pioneer was American chemist Percy Lavon Julian (1899–1975). Early in his career he invented physostigmine, a drug that could successfully treat glaucoma, an eye disease. Thousands of other drugs are now in use worldwide. They include antiseptics (to kill bacteria), anesthetics (for dulling pain), cytotoxic drugs (for killing cancer cells), and anti-inflammatory drugs (which can relieve the swollen, painful joints of arthritis).

During the 20th century, drugs became an especially important treatment for mental illnesses. In the early 1950s, Czech-born chemist Frank Berger (1913–) discovered chemicals that could calm angry monkeys—and thus invented the first human tranquilizer drugs, Miltown and Equanil. Shortly afterward, chemist Leo Sternbach (1908–2005), working for the Roche company, developed tranquilizers with fewer side effects. In the following decade, Ray Fuller (1935–1996) and his team at the Eli Lilly Company began selling fluoxetine, a drug that proved to be an effective treatment for depression. Popularly known as Prozac, it became the most popular psychiatric drug in history within three years of its launch.

Apart from bringing real benefits to health, effective drugs can tell medical scientists a great deal about how the body works. For example, scientists know that Prozac affects the way the chemical serotonin is absorbed by the brain. Since Prozac can relieve depression, its effectiveness has helped to confirm that serotonin is involved in depressive illnesses. Successful drugs therefore play an important part in increasing medical knowledge.

ideas propounded by the ancients proved hugely influential. Galen's work was used for centuries, and Hippocrates gave his name to the "Hippocratic Oath," a pledge to put the interests of patients first that still guides physicians today.

INVENTING THE MEDICAL REVOLUTION

Medicine moved onto a more scientific footing during the Renaissance, around the fifteenth century, when science began to challenge religion as the most accurate explanation of human experience. In the 1490s, the famous Italian artist and inventor Leonardo da Vinci (1452–1519) made sketches of the body's inner workings in his notebooks. About fifty years later, Flemish physician Andreas Vesalius (1514–1564) published the first reliable textbook of detailed drawings showing the inner structures of the human body.

Scientific inventions helped medicine greatly. During the seventeenth century, English scientist Robert Hooke (1635–1703) developed the modern compound microscope (one with several lenses), which made possible the study of organisms invisible to the naked eye. Using his invention, he observed the tiny "building blocks" from which living things were made and became the first person to call them cells. Used to study cadavers, microscopes greatly expanded the sciences of anatomy and physiology. For example, later medical scientists, such as the German Rudolf Virchow (1821–1902), built on Hooke's work, showing how living cells could make healthy tissue grow but, at the same time, were at the center of illness and disease.

About fifty years later, German physicist Daniel Fahrenheit (1686–1736) invented the first practical thermometers and the Fahrenheit temperature scale. This enabled physicians to monitor people's health by measuring how hot or cold their bodies were. Another great medical invention came about a century later, in 1819, when French physician René Laënnec (1781–1826) invented the stethoscope, an instrument that helps a physician listen to sounds inside a person's chest. The thermometer and stethoscope gave physicians a better idea of what might be happening in someone's body. They proved to be great aids to diagnosis (finding out a person's illness), which is the first step to successful treatment.

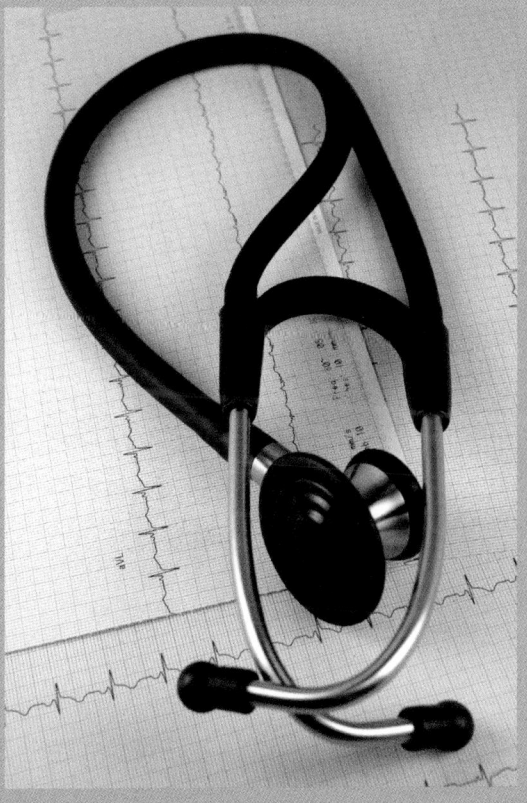

The stethoscope, invented by physician René Laënnec in 1819.

Illustration from around 1820 of a smallpox ward in Hampstead Hospital, England (artist unknown).

PREVENTING DISEASE

One of the greatest medical advances of the 18th century came from an inspired discovery made by English physician Edward Jenner (1749–1823). At that time, many people were dying from smallpox. Jenner noticed that people suffering from another disease, cowpox, did not get smallpox. This led Jenner to invent vaccination, a treatment in which people are given a minute dose of the virus that causes a serious disease to prevent them from catching it later. Jenner also studied the causes of smallpox and was the first to use the word *virus* to explain how the disease was transmitted.

At the time the causes of disease were hotly debated. Some physicians believed in an old theory—spontaneous generation—that claimed diseases could arise from inanimate objects. About a half-century after Jenner's breakthrough, Louis Pasteur (1822–1895), a French scientist, proved that diseases are actually spread by bacteria and viruses; this idea became known as the germ theory of disease. Using this theory, Pasteur developed important vaccines for diseases such as anthrax, cholera, and rabies. His scientific breakthrough also enabled him to invent pasteurization, a way of preserving foods by heating them briefly to kill germs.

Pasteur's work began the science of bacteriology, which was in turn advanced by others. A Scottish surgeon, Joseph Lister (1827–1912), realized that germs in the operating room greatly endangered the lives of his patients. When he started spraying the operating theater with the antiseptic (germ-killing) carbolic acid before operations, many more of his patients survived. In the 1880s, German physician Robert Koch (1843–1910) took Pasteur's work in a different direction, using microscopes to identify the bacteria that caused diseases such as tuberculosis and cholera.

Penicillin, discovered by Alexander Fleming in 1928, being mass-produced in a factory in Liverpool, England, in 1954.

Another important breakthrough in the fight against disease came in 1928, when the English bacteriologist Alexander Fleming (1881–1955) accidentally invented penicillin. Fleming found that a mold growing in a dish in his laboratory was very effective at killing bacteria. Later, the mold (*Penicillium notatum*) was used to make the antibiotic (bacteria-killing) drug penicillin. Many other antibiotics have been developed since then.

ADVANCES IN SURGERY

Surgery, the process of cutting into a person's body to repair damage or remove diseased tissue, has been used by doctors since ancient times. Until the 19th century, however, surgery was often agonizing because patients were usually conscious throughout their operation. A great development in surgery came in 1846 when the American dentist William Morton (1819–1868) invented general anesthetic, a way of using a gas to render the patient unconscious; thus the patient did not feel the pain of surgery. Almost exactly a century later, American physician Charles Drew (1904–1950) made another great surgical advance when he set up the world's first large-scale blood banks (stores of donated blood).

Advances such as these made possible ever more ambitious—and contentious—operations. One of the most controversial forms of surgery yet attempted was pioneered in 1967 when a South African doctor, Christiaan Barnard (1922–2001), carried out the world's first heart transplant. Such dramatic operations carry a high risk of failure and are immensely traumatic for patients.

Minor operations are now often carried out in a less intrusive way using the keyhole surgery technique—a small incision is made in a patient's body and the procedure is done without making large incisions. Lasers (pure, high-powered light beams) are also used in modern operations because they can cut soft tissue very precisely. They were invented in the 1950s by American physicists Arthur Schawlow (1921–1999) and Charles Townes (1915–). Lasers have been widely used in eye surgery since the 1980s, when African American physician Patricia Bath (1942–) pioneered their use for removing cataracts, a cloudiness in the lens of the eye that can cause blindness.

A patient undergoes an x-ray procedure in 1942.

ATOMIC MEDICINE

The most dramatic scientific and technological advances of the 20th century came about with the discovery of the world inside atoms, the tiny particles from which all substances are made. More effective medical diagnosis and treatment were among the more positive developments of atomic technology. X-rays (powerful, invisible electromagnetic waves) were accidentally discovered in 1895 by German physicist Wilhelm Röntgen (1845–1923) and have been used by physicians ever since to study such things as broken bones.

In the 1970s, British scientists Allan Cormack (1924–1998) and Godfrey Hounsfield (1919–2004) developed a type of improved x-ray machine, known as computerized axial tomography (CAT or CT). As its name suggests, it used a computer to make detailed, three-dimensional images of the body by scanning x-ray beams through it. Medical techniques like this have revolutionized diagnosis, especially in neuroscience (the study of the brain). Another revolutionary form of

TIME LINE

460–377 BCE	130 BCE–200 CE	1490s	1540s	1660s	1714
Hippocrates begins his inquiries into disease.	Galen founds the study of anatomy and physiology.	Leonardo da Vinci sketches the body's inner workings.	Andreas Vesalius publishes the first anatomy textbook.	Robert Hooke develops the compound microscope.	Daniel Fahrenheit invents the first practical thermometers.

784 HEALTH AND MEDICINE

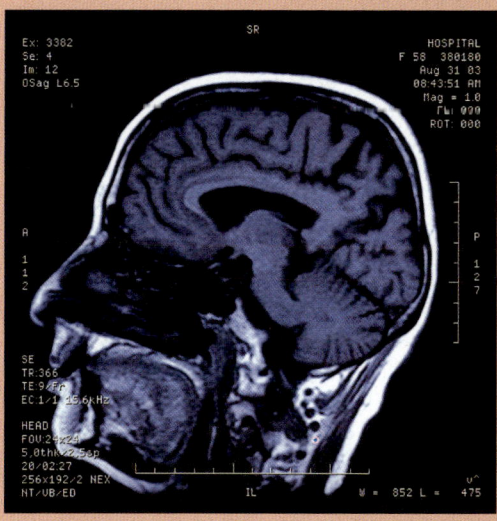

An MRI of the human brain.

scanning known as magnetic resonance imaging (MRI) was developed in the 1970s and 1980s by biological physicist Raymond Damadian (1936–). Unlike x-rays, which show hard substances like bones and teeth, MRI builds a picture of the soft tissues in the body and is better for locating problems such as tumors.

Atomic technology has proved equally useful in treating some of the illnesses it has helped to diagnose. One year after Röntgen chanced upon x-rays, French physicist Antoine-Henri Becquerel (1852–1908) discovered radioactivity (high-energy rays or tiny particles given off by unstable atoms). Another pioneer in this area, Polish-born physicist and chemist Marie Curie (1867–1934), used her knowledge to discover a new chemical element, radium, in 1898. Soon afterward, radium was used to develop the first radiation treatments for cancer—the disease that, tragically, claimed Curie's own life.

GENETICS

Even before the advent of modern medicine, physicians had observed that certain illnesses tended to run in families. Since the mid-20th century, a new science—genetics—has given doctors an entirely new insight into inherited disease. The genetics age began in 1953 when English physicist Francis Crick (1916–2004) and American biologist James Watson (1928–) discovered the structure of the chemical deoxyribonucleic acid (DNA). DNA is contained in every living cell and carries the genetic information that works like a set of instructions to tell the cell how to develop.

Since the discovery of DNA, biologists have believed that DNA also contained the secrets of many illnesses because it could instruct cells to malfunction. By studying DNA, scientists hoped

TIME LINE

1796	1819	1846	1850s	1860s	Late 1860s
Edward Jenner conducts his first successful smallpox vaccination.	René Laënnec invents the stethoscope.	William Morton develops general anesthetic.	Louis Pasteur invents pasteurization.	Pasteur develops the germ theory of disease.	Joseph Lister begins using antiseptic during surgery.

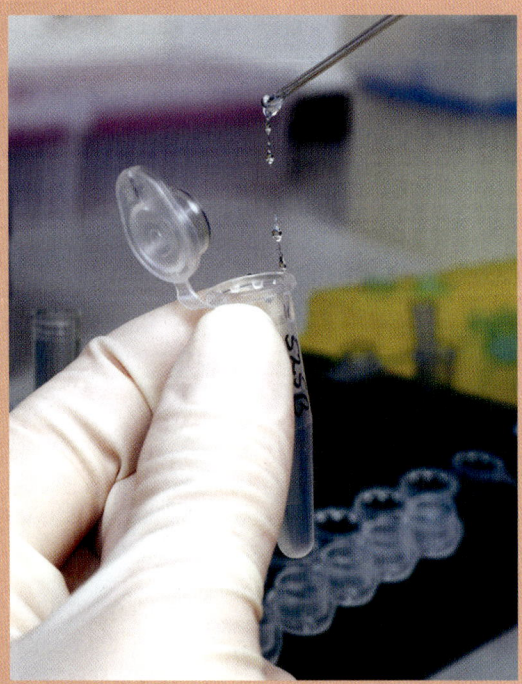

In 2000, at the Great Ormond Street Hospital in London, a molecular geneticist spools DNA out of a solution at the end of a DNA extraction.

to discover how to stop certain illnesses from developing or prevent them from being passed from parents to children. Before scientists could begin to do so, however, they had to discover the meaning of the body's entire set of genetic information, which is known as the human genome. An ambitious 13-year worldwide effort, the Human Genome Project, had successfully mapped all of the information in human DNA (equivalent to 200 volumes of a typical telephone directory) by 2003. However, scientists are still a long way from knowing what all this information means or how to use it to cure disease.

The process of manipulating genetic information is called genetic engineering. Removing all or part of an organism's DNA and transferring it into another organism in which it can grow are a means of exactly copying living cells. This technique, cloning, was developed in the early 1970s by Herbert Boyer (1936–) and Stanley Cohen (1935–). Its first use was in manufacturing the drug insulin (a treatment for diabetes). One day, cloning might be used to grow replacement body parts or even entire human beings—though its use in such areas would be highly controversial.

MODERN MEDICAL INVENTIONS

Great improvements to people's lives are often achieved through simple inventions. One example has been the development of fiber optics: a way of transmitting light down flexible glass pipes that has revolutionized telecommunications. The idea was originally developed in the 1950s by

TIME LINE (continued)

1880s	Late 1800s	1895	1896	1897	1898
Robert Koch identifies tuberculosis and cholera bacteria.	Rudolf Virchow develops the field of pathology.	Wilhelm Röntgen discovers x-rays.	Antoine-Henri Becquerel discovers radioactivity.	Felix Hoffmann makes aspirin.	Marie Curie discovers radium.

Medical Materials

One of the biggest areas of technological advances of the 20th century was the development of artificial (synthetic) materials, which have revolutionized everything from clothing design to aerospace. Medicine has also benefited enormously from these new materials. For example, nylon, the revolutionary synthetic fiber invented in the 1930s by American chemist Wallace Carothers (1896–1937), is one of the materials used to stitch wounds together.

Since the 1970s, surgeons have been replacing worn-out body parts with bionic (artificially engineered) replacements made from such materials as silicone, rubber, plastics, and carbon fibers. One of the best-known examples is used in hip replacement surgery. In place of worn-out bone joints, many older people—especially those suffering from arthritis—now have artificial hip joints made from metal balls that rotate smoothly in plastic sockets. The latest artificial limbs (prosthetics) are also made from high-tech materials such as carbon fibers.

Another important treatment involves growing new tissues in the laboratory, then using them to repair damaged or diseased parts of the body. If a person suffers skin damage, perhaps in a fire, a small sample of existing skin, the size of a postage stamp, can be used to grow a large area of replacement skin (skin autograft) in just a few weeks.

A related treatment involves using fetal tissue (tissue removed from human embryos or aborted fetuses) to replace or regenerate damaged tissue in a patient's body or brain. For example, doctors have managed to relieve the symptoms of Parkinson's disease (a serious illness afflicting the human nervous system) by injecting fetal cells into the basal ganglia—a part of the brain. Although the treatment is effective, it has been very controversial and raises ethical issues about the uses of human embryos.

TIME LINE

1920s	1935	1940s	1941–1945	Early 1950s	Late 1950s
Alexander Fleming discovers penicillin.	Percy Lavon Julian develops physostigmine.	Charles Drew sets up the first large-scale blood banks.	Bessie Blount develops the automatic feeding apparatus for the disabled.	Narinder Kapany develops fiber optics.	Arthur Schawlow and Charles Townes develop lasers.

Indian physicist Narinder Kapany (1927–). Later in that decade, American scientists used fiber optics to build the first gastroscope, a medical device that could help doctors see inside people's stomachs by looking down a flexible glass tube inserted down a patient's throat.

In vitro fertilization (IVF) is another simple invention that has made a huge difference to many people who have been unable to conceive children naturally. IVF involves removing eggs from a woman's ovaries and fertilizing them with a man's sperm in a laboratory dish. The fertilized eggs are then returned to the woman's womb, where they develop normally. The technique was pioneered by two British doctors, Patrick Steptoe (1913–1988) and Robert Edwards (1925–), who helped to produce the first IVF baby, Louise Brown, in July 1978. In 2006 researchers estimated that more than 3 million babies worldwide had been born by IVF.

Many inventors have made a difference to the lives of the disabled. An invention developed by American Bessie Blount (1914–) greatly improved the lives of people who had lost the ability to feed themselves. Blount's idea was an electrically controlled feeding tube that could deliver portions of food automatically whenever the disabled person bit into it.

Another American inventor, Dean Kamen (1951–), is perhaps best known for developing an electrically powered, two-wheeled trolley—the Segway personal transporter. However, he has also made inventions for the disabled, including the revolutionary iBOT wheelchair. With four-wheel drive and a built-in gyroscope (heavy, rapidly spinning wheel) to aid in balance, the iBOT can negotiate rough terrain, lift itself up to standing height, and even climb stairs. Kamen's other inventions include a mobile dialysis machine (to clean the blood of someone with kidney dysfunction), an automatic device for giving insulin injections to people with diabetes, and an engine that can purify water for developing countries.

Lack of clean water is one of the biggest causes of illness in the developing world. Figures published in 2002 by the World Health Organization show that, worldwide, 1.1 billion people lack basic water facilities and 2.6 billion people do not have proper sanitation. As a result, an estimated 3.4 million people (mostly children under age five) die each year from illnesses such as diarrhea, cholera, and hepatitis, which can be carried in dirty water. Kamen is not the only inventor to try to tackle this problem. In 1993, Indian-born American scientist Ashok Gadgil (1950–)

TIME LINE (continued)

1950s	1953	1960s	1967	1971	Early 1970s
Frank Berger invents human tranquilizer drugs.	Francis Crick and James Watson discover DNA.	Ray Fuller develops fluoxetine (Prozac).	Christiaan Barnard carries out the world's first heart transplant.	Dean Kamen invents the AutoSyringe.	Stanley Cohen and Herbert Boyer develop cloning.

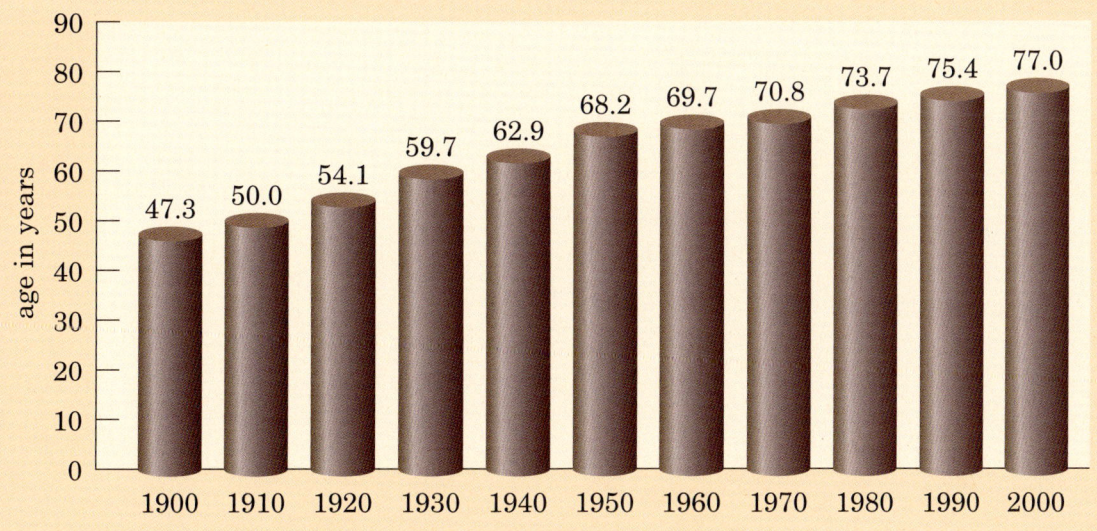

Advances in medicine are one reason for the dramatic increase in U.S. life expectancy rates in the 20th century.

began developing a simple invention that could disinfect a ton of water (.9 metric ton) for just half a cent. Using ultraviolet light, and powered by a simple car battery, it can kill 99.999 percent of bacteria and produce enough clean water for 1,000 people.

How many millions of lives have medical inventions helped save? This question is impossible to answer. Some of the most effective inventions, including devices to clean water supplies and vaccination, greatly reduce the chances of people becoming ill; others, such as blood banks and antibiotics, improve people's chances of recovery when they are already unwell. Microscopes and other scientific instru-

TIME LINE

1970s	1970s–1980s	1978	1980s	1993	2006
Allan MacLeod Cormack and Godfrey Newbold Hounsfield develop CAT.	Raymond Damadian develops MRI.	Louise Brown, the first IVF baby, is born.	Patricia Bath pioneers the use of lasers in cataract removal.	Ashok Gadgil begins to develop the UV Waterworks system.	More than 3 million babies are estimated to have been born by IVF.

ments have greatly advanced medical science and knowledge; stethoscopes, x-rays, and the latest medical scanning technologies have made diagnosing illness more accurate and treating illness easier. Advances such as these have made an enormous difference to human health and well-being.

—Chris Woodford

Further Reading

Books

Cule, John. *Timetables of Medicine*. New York: Black Dog & Leventhal, 2000.

Drugs and Society. New York: Marshall Cavendish, 2006.

Parker, Steve. *Eyewitness: Medicine*. New York: Dorling Kindersley, 2001.

Porter, Roy. *Blood and Guts: A Short History of Medicine*. New York: Penguin, 2003.

Tiner, John Hudson. *Exploring the History of Medicine*. Green Forest, AR: New Leaf/Master, 1999.

Web sites

Body and Mind
An educational site about medicine and health.
http://www.bam.gov/

Human Genome Project
Information about the history, science, and benefits of the Human Genome Project.
http://www.ornl.gov/sci/techresources/Human_Genome/home.shtml

KidsHealth
Information, activities, and advice on health issues.
http://www.teenshealth.org/teen/

Virtual Knee Surgery
Take on the role of the surgeon throughout a total knee replacement surgery.
http://www.edheads.org/activities/knee/

See also: Bath, Patricia; Blount, Bessie; Boyer, Herbert, and Stanley Cohen; Carothers, Wallace; Cormack, Allan, and Godfrey Hounsfield; Damadian, Raymond; Drew, Charles; Edwards, Robert, and Patrick Steptoe; Fahrenheit, Daniel; Gadgil, Ashok; Hooke, Robert; Jenner, Edward; Julian, Percy Lavon; Kamen, Dean; Kapany, Narinder; Morton, William; Pasteur, Louis; Schawlow, Arthur, and Charles Townes.

BEULAH HENRY

Inventor of household devices
1887–1973

Beulah Henry may not be a famous name today, but she was one of the most prolific female inventors of household and business products during the 1920s, 1930s, and 1940s. In all, Henry was responsible for more than one hundred inventions. She held 49 patents and was nicknamed "Lady Edison." Henry is considered to be one of the most inventive women in the United States.

EARLY YEARS

Although Henry was a prolific inventor, not much is known about her life. Born in Memphis, Tennessee, in 1887, Beulah Louise Henry is believed to be a descendent of Patrick Henry, the American revolutionary who cried, "Give me liberty or give me death!"

Some sources suggest that Henry's knack for invention began early on, when she was a young girl who sketched her creations. Art was a family interest—her father was an authority on art, her mother was an artist, and her brother was a songwriter. Records show that she attended Presbyterian and Elizabeth colleges, in Charlotte, North Carolina, graduating in 1909.

FIRST INVENTIONS

In 1912 at age 25, while still living in North Carolina, Henry received her first patent, for an ice cream freezer with a vacuum seal. The following year, she patented a handbag and parasol. Her early success prompted her to move to New York City.

In New York, Henry created one of her best-known inventions, an umbrella with a detachable snap-on cloth cover that permitted women to coordinate their rain gear with different outfits. When Henry first

Undated portrait of Beulah Henry.

suggested the idea, umbrella manufacturers of the day told her it could not be done. She forged ahead despite their skepticism and created the product in 1924. Her umbrellas would earn her $50,000—a considerable sum in the 1920s—and a prominent spot in the windows of Lord and Taylor, one of New York City's leading department stores. On the basis of the umbrella's success, she founded the Henry Umbrella and Parasol Company.

During the 1920s, Henry also invented various products related to sponges, including the Latho, which was a sponge with a special compartment that opened and snapped shut to hold a bar of soap inside. The product also floated, to help bathers locate the soap in the bath. Manufacturers proved unable to cut sponges to produce the Latho correctly, so Henry developed a machine that could. In 1929, she patented "Dolly Dips," the children's version of the Latho.

By age 40, Henry had founded her second company, the B. L. Henry Company of New York, which produced a variety of her inventions. She was one of the few of that era to have inventions patented in four different countries and to have two businesses in her name.

PLAY TIME

Another domain of Henry's inventing was toys, including several types of dolls. The most famous, perhaps, was the "Miss Illusion" doll, from 1935. "Miss Illusion" had interchangeable wigs—one blonde, one brunette—and a reversible dress. The doll's eyes, which opened and closed by a mechanism inside the head, were also changeable, from blue to brown. Henry invented dolls with spring-loaded limbs and bendable arms, dolls that could kick, blink, eat, and even talk. Another doll had a radio inside.

Henry also produced educational products for children, including "Kiddie Klock," which taught children to tell time; and "Cross Country," a board game that taught children about U.S. geography. Unlike her other inventions, many of these toys were never patented and were simply sold to various toy companies.

MACHINES

During the 1930s and 1940s, Henry shifted her focus to machines—most notably, the typewriter. Invented in 1867 by Christopher Latham Sholes (1819–1890), the typewriter revolutionized communication and became one of the most common implements in everyday business. Between 1932 and 1964, Henry received 11 patents related to the typewriter. Her first was for the "protograph," a machine that worked with a manual typewriter to create several copies of a single document without

TIME LINE

1887	1912	1924	1927
Beulah Louise Henry born in Memphis, Tennessee.	Henry receives her first patent.	Henry creates a new type of umbrella.	Henry founds the B. L. Henry Company of New York.

using carbon paper. During World War II, with carbon shortages, the protograph proved very useful. Other inventions included devices for feeding documents into and aligning documents for typewriters and duplicating machines. In 1937, she invented a cash register that could write like a typewriter for the National Cash Register Company.

In 1939, Henry was hired as an inventor by Nicholas Machine Works. The company provided her with a lab and technical staff. While there, she was credited with several dozen inventions, some of which, as had happened with the Latho, required her to develop not only the product but also the means of production.

A typewriter from the 1920s outfitted with a protograph, which enabled the typist to make several copies of a document.

TIME LINE

1932–1964	1939	1970	1973
Henry receives 11 patents for inventions related to the typewriter.	Henry hired as an inventor by the Nicholas Machine Works.	Henry receives her final patents.	Henry dies.

THE LADY EDISON

During her amazing career, Henry apparently never married. She spent much of her life living in hotels, in particular New York's Hotel Victoria. Records show she was a member of many city establishments, including the Museum of Natural History, the Audubon Society, and the League for Animals. Various mentions in the press suggest that Henry enjoyed a somewhat public, if enigmatic, persona. Indeed, news stories mentioned her "superb auburn hair" and her "commanding presence." She told

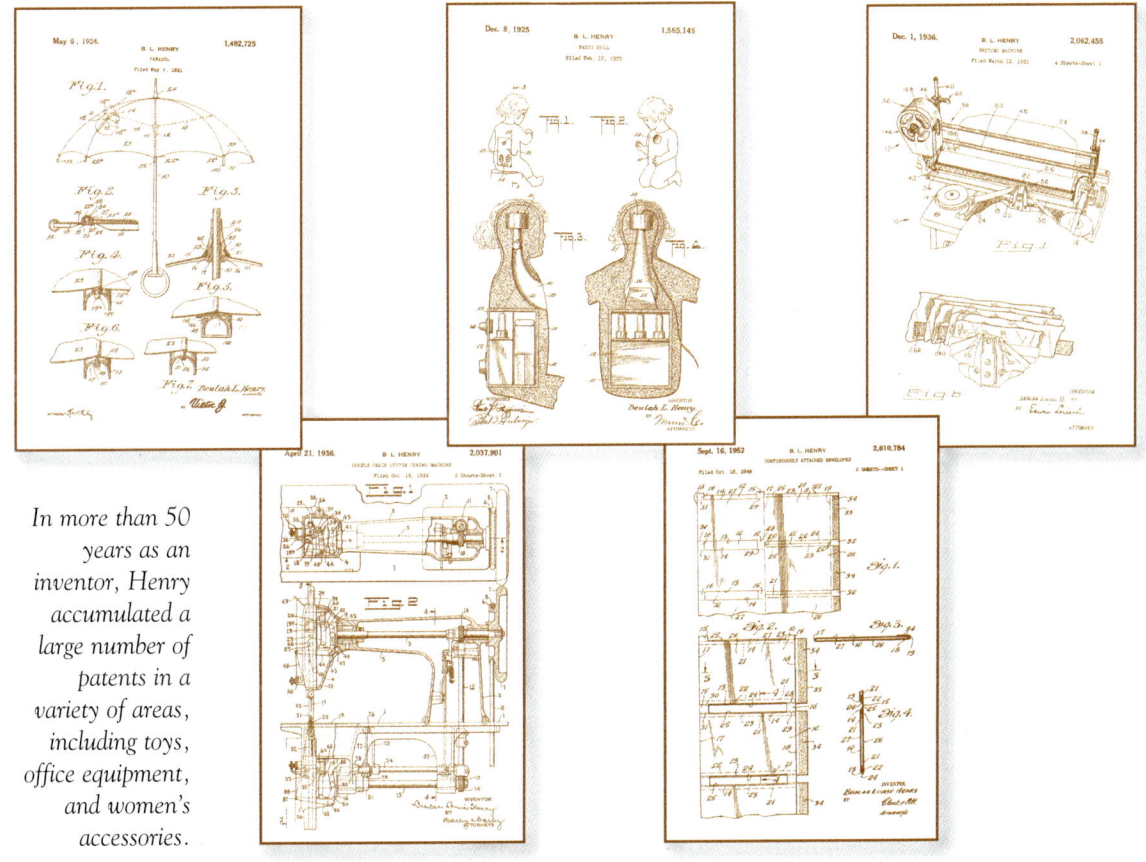

In more than 50 years as an inventor, Henry accumulated a large number of patents in a variety of areas, including toys, office equipment, and women's accessories.

reporters of her fondness for painting. She continued to invent well into old age, receiving her final patent in 1970, at the age of 83. She died three years later.

What is most significant about Henry's career, aside from the sheer number of her inventions, was that, unlike most women of her day, she profited significantly from her work. Many inventions by women in that era were credited to their husbands—at that time, inventions by women accounted for less than 1 percent of all U.S. patents.

Henry is still remembered as "Lady Edison," a nickname bestowed upon her by the U.S. Patent Office. Some, in particular the women engineers who have helped keep Beulah Henry in the history books, find the appellation condescending, because Henry was clearly an outstanding inventor in her own right.

—Laura Lambert

Further Reading

Casey, Susan. *Women Invent: Two Centuries of Discoveries That Have Shaped Our World.* Chicago: Chicago Review, 1997.

Stanley, Autumn. *Mothers and Daughters of Invention.* Piscataway, NJ: Rutgers University Press, 1995.

See also: Communications; History of Invention; Household Inventions; Sholes, Christopher Latham.

HISTORY OF INVENTION

Archaeologists have found that the earliest human inventions were objects such as simple tools made from stone and wood. Fire, believed to have been harnessed for the first time between one and two million years ago, was another early human discovery. Compared with modern machines such as earthmovers and space rockets, these early developments seem primitive. Yet virtually every modern invention has evolved from one or more ancient ones: a 21st-century bulldozer uses the same basic physics as a prehistoric stone ax, and no modern space rocket could take off without fire. The story of technology begins with prehistoric inventions, which laid the foundations for all inventions that followed.

ANCIENT TIMES

People first started to settle in fixed places between ten and eleven thousand years ago. The first civilizations were most likely in the Middle East—in Egypt as well as in an area called Mesopotamia (now Iraq, Syria, and Turkey). In this area some of the most important inventions of all time were developed, such as agriculture, written language, and the wheel. In a time when so few inventions existed, every new discovery had the potential to make an enormous difference to humankind.

The development of written language by the Sumerians around 3500 BCE dramatically changed the process of invention. Written language allowed the recording and passing on of information, which aided the development of such areas of

Pointed harpoons on display at the National Museum of Prehistory in France; dating from around ten thousand years ago, these tools signify the development of new hunting skills.

study as history, education, and science. The development of science was particularly critical in helping early people understand why the world worked the way it did and led to the invention of new technologies. The ancient Greeks, whose society was at its zenith two to three thousand years ago, were among the world's first scientists. They invented many new technologies using mathematical ideas and a type of careful thinking known as philosophy. The Greek philosopher and mathematician Archimedes (ca. 287–212 BCE), for example, greatly advanced scientific knowledge and devised simple machines, such as levers and pulleys.

Ancient peoples became adept at taking ideas from earlier civilizations and improving upon them. Using this process, the ancient Egyptians were able to advance boat-building technology. Earlier peoples had made dugout canoes by hollowing out a single, large tree. The Egyptians, however, used a dugout as a base and built planks upward from its sides, which created a boat of potentially unlimited size. The Romans, whose empire prospered from around 29 BCE to 495 CE, also further developed earlier ideas, in particular those of the ancient Greeks. Whereas the Greeks invented gears (wheels with teeth around their edges that magnify the speed or force of a machine), the Romans were the first to employ and deploy them on a large scale.

Ancient peoples had invented written languages, with letters and words that they could put together in infinite combinations to express themselves. Now they also had a small collection of inventions that could be combined or used in different ways to make entirely new inventions. Wheels that had been invented to help move loads from place to place were by the first century CE being used in waterwheels and windmills, both of which generated power. When the Romans combined wheels with levers, they invented the crane.

THE MIDDLE AGES

The period of history that lasted roughly from the end of the Roman Empire to the 1500s is often known as the Middle Ages (or medieval period). During this time a number of important new inventions were introduced. One of the most important, gunpowder, was invented in China between 700 and 900 CE. Another military breakthrough during the Middle Ages

In part a tribute to Archimedes, Theatrum Machinarum Novum *was published by Georg Bockler in 1671. This engraving from the book shows a grinding mill (M) driven by a water wheel (H) and an Archimedes screw (Q).*

was the development of a harness that allowed warriors to fight on horseback, from where they could easily defeat foot soldiers. Transportation also saw advances: the Chinese invented the magnetic compass around the eleventh century and it soon spread to Europe, revolutionizing navigation at sea.

During the early Middle Ages, science, mathematics, and technology progressed most rapidly in the Islamic world. Ideas slowly filtered back to Europe from about the twelfth century onward. In the fourteenth, fifteenth, and sixteenth centuries, European creativity in science, technology, art, and invention moved forward at an amazing pace—this time became known as the Renaissance ("rebirth"). No one typifies this era better than Leonardo da Vinci (1452–1519), the highly creative Italian who excelled as an artist, scientist, and inventor. One of the most important devel-

Inventing Myths

The history of invention conceals many myths. Most people believe Alexander Graham Bell (1847–1922) invented the telephone, yet the truth is more complex: several others developed similar inventions around the same time and it can be considered largely an accident of history that Bell is the only one people remember. Guglielmo Marconi (1874–1937) is known as the inventor of radio, but others had discovered the basic science before him; Marconi's genius was not so much in the invention of radio as in turning it into a practical technology that captured people's imagination. Many history books portray Louis Pasteur (1822–1895) as a brilliant scientist who developed important vaccines and a revolutionary technique for preserving food (pasteurization), but modern biographers claim some of his work was scientifically dubious and suggest that he borrowed some of his best discoveries from others without giving due credit.

Many people also have misconceptions about the process of inventing, believing that inventions arrive in the world by sudden flashes of inspiration. Indeed, some inventions do appear this way; Velcro was invented after George de Mestral (1907–1990) found burrs stuck to his pants after a walk in the mountains and realized he could use the same principle to make a loop-and-hook fastener for clothing. In contrast, Thomas Edison (1847–1931) filled forty thousand pages of notes and experimented with more than six thousand different materials until he finally hit upon the perfect design for the lightbulb.

An engraving from the fourteenth century shows the devil on the shoulder of a monk involved in an experiment—an allegorical representation of the invention of gunpowder.

opments during the Renaissance was the arrival of gunpowder in Europe. Manufactured on a large scale from the fourteenth century onward, it revolutionized warfare because it enabled armies to fight at a greater distance from their enemies and made huge battles and wars possible. Gunpowder has changed the course of history many times because of the various weapons and explosives that have been devised to use it.

Another invention of the Renaissance, the modern printing press, affected history no less dramatically. Developed in the mid-fifteenth century by German metalworker Johannes Gutenberg (ca. 1400–1468), who incorporated some earlier printing technologies originating in China, the press enabled a vast dissemination of information and knowledge. Before Gutenberg's printing press, books were written (copied) individually by hand; following his invention, books could be printed inexpensively in huge quantities. Printed books revolutionized education, allowing history, science, and other kinds of knowledge to circulate widely.

THE SCIENTIFIC REVOLUTION

Science took on greater importance during the 1600s and 1700s, when it began to replace religion as the most credible explanation for phenomena of the everyday world. This period of history—known as the scientific revolution—was brought about partly by new inventions.

Microscopes, pioneered by Antoni van Leeuwenhoek (1632–1723) and Robert Hooke (1635–1703), opened up an invisible world of living microorganisms and soon began to revolutionize the life sciences, including medicine. Telescopes, developed by Italian physicist Galileo Galilei (1564–1642), helped to prove that the sun was the most important object in the earth's solar system.

Inventions such as these enabled scientists to see how things worked as they did and to understand the governing principles of the physical world; these had previously been a mystery. Once Daniel Fahrenheit (1686–1736) invented the first practical thermometers and his Fahrenheit temperature scale early in the 18th century, later pioneers were able to articulate advanced scientific theories of heat energy. These theories, in turn, played a crucial role in the development of steam and gasoline engines in the centuries that followed.

THE INDUSTRIAL REVOLUTION

Steam engines first appeared during the Industrial Revolution, a period of history spanning the 18th and 19th centuries when many people left the countryside and the work of farming for jobs in industrialized towns and cities. The Industrial Revolution began as an agricultural revolution in the early 1700s when English farmer Jethro Tull (1674–1741) invented machines that could sow seeds and plow fields with less effort. As farms became increasingly mechanized, so, too, did the products of farming. In 18th-century England, James Hargreaves (ca. 1720–1778) developed a machine that automated the production of textiles. These inventions led to a huge expansion in cloth production, but also caused riots as angry workers, such as a group called the Luddites, smashed power looms to protest the possible loss of their jobs.

In the United States, Eli Whitney (1765–1825) developed the cotton gin, which cleaned harvested cotton, making it ready to be spun into thread. The cotton gin greatly increased agricultural production and brought wealth to many in the southern United States. One unforeseen effect of increased cotton production was a major expansion of the slave trade (workers were needed to harvest the crop), which contributed to the outbreak of the American Civil War.

TIME LINE

8,000–9,000 BCE	3500 BCE	ca. 287–212 BCE	29 BCE–495	100s CE
Humans begin to form permanent settlements.	Sumerians develop a written language.	Archimedes advances scientific knowledge and devises simple machines.	Romans advance inventions and ideas of the Greeks.	Wheels are used in waterwheels and windmills to generate power.

A reconstruction of the workshop of engineer James Watt.

Although the cotton gin was one of the first inventions to be granted a U.S. patent under the 1790 Patent Act, Whitney's invention was so easy to copy that the patent was of little use—a foreshadowing of the difficulties some later inventors would face trying to protect their ideas. Ironically, Whitney had designed the cotton gin (as well as other inventions) so it could be mass-produced in factories by relatively unskilled workers. This approach was a significant departure from traditional methods of manufacture, which employed skilled artisans to make objects by hand, one at a time.

The Industrial Revolution accelerated during the 18th century, after English inventor Thomas Newcomen (1663–1729) developed the first practical steam engine. Originally used for pumping wastewater out of mines, steam engines were soon powering looms and other machines in textile factories. Later inventors, notably James Watt (1736–1819), a Scot, refined Newcomen's invention to make it smaller, more powerful, and portable. These compact steam engines were then used to power ships pioneered by Robert Fulton (1765–1815), railroad locomotives developed by George Stephenson (1781–1848), and even an ill-fated airplane invented by Samuel P. Langley (1834–1906).

TIME LINE (continued)

700–900	1000s	Middle Ages	1400s	1609	1660s
Gunpowder is invented in China.	The Chinese invent the magnetic compass.	A harness allowing warriors to fight on horseback is invented.	Johannes Gutenberg develops the modern printing press.	Galileo Galilei develops the telescope.	Robert Hooke develops the compound microscope.

Twenty-first-century developments in wireless computing allow bedouin to go online even in the deserts of Tunisia.

MODERN TIMES

Like their predecessors, most modern technologies have their roots in earlier discoveries. Electric power, synthetic materials, the electronic computer, and the Internet are some inventions that define modern times. Electric power, which was pioneered in the late 19th century by Thomas Edison (1847–1931), can be traced to the ancient Greeks, who first discovered electricity. Nylon, the first significant synthetic fabric, was developed by the American chemist Wallace Carothers (1896–1937) of the DuPont Company. The manufacture of nylon uses spinning

TIME LINE

Early 1700s	1710s	1714	1760s	1790s	Early 1800s
Jethro Tull develops first farm machines.	Thomas Newcomen invents the steam engine.	Daniel Fahrenheit invents the first practical thermometer.	James Hargreaves develops the spinning jenny.	Eli Whitney develops the cotton gin.	Robert Fulton develops the steamboat.

techniques similar to those in use even before the arrival of automated textile machines in the 18th century. In much the same way, computers can be traced through several thousand years of history back to the abacus; and the Internet owes its existence to two 19th-century methods of long-distance communication: the telegraph, developed by Samuel Morse (1791–1872), and the telephone, whose invention (though disputed) is usually credited to Alexander Graham Bell (1847–1922).

Most modern inventions are developed not by lone inventors but by large corporations, where many scientists and engineers work together. This process of corporate invention became popular in the United States around World War II and has continued ever since. Modern inventions define the world people live in today—a world of automated industry, rapid travel, high technology, and instant communications. One example is the trend known as globalization—the idea that jobs, goods, services, money, and ideas can move freely around the world. Globalization did not just happen spontaneously—many inventions made it possible, including the telegraph and telephone, container shipping, jet airplanes, and the World Wide Web, which was invented by Tim Berners-Lee (1955–) in the late 1980s.

—Chris Woodford

Further Reading

Bender, Lionel. *Eyewitness: Invention*. New York: Dorling Kindersley, 2005.

Bridgman, Roger. *1,000 Inventions and Discoveries*. New York: Dorling Kindersley, 2006.

Toffler, Alvin. *Future Shock*. New York: Bantam Books, 1991.

Williams, Trevor. *A History of Invention*. New York: Time Warner, 2004.

See also: Archimedes of Syracuse; Bell, Alexander Graham; Berners-Lee, Tim; Carothers, Wallace; De Mestral, George; Edison, Thomas; Fahrenheit, Daniel; Fermi, Enrico; Ford, Henry; Fulton, Robert; Galilei, Galileo; Goddard, Robert H.; Gutenberg, Johannes; Hargreaves, James; Hooke, Robert; Marconi, Guglielmo; Morse, Samuel; Newcomen, Thomas; Pasteur, Louis; Stephenson, George; Whitney, Eli.

TIME LINE (continued)

1814	1830s–1840s	1870s	1880s	1930s	Late 1980s
George Stephenson builds the first railroad locomotive.	Samuel Morse develops the telegraph.	Alexander Graham Bell develops the telephone.	Thomas Edison pioneers electric power.	Wallace Carothers invents nylon.	Tim Berners-Lee invents the World Wide Web.

SOICHIRO HONDA

Inventor of the small
motorcycle engine

1906–1991

Perceiving a need for light transportation in postwar Japan, Soichiro Honda attached a small engine to a bicycle, and on the strength of this idea he eventually turned his small machine shop into a global corporation. Honda brought an innovative approach to his inventing, using technology from high-performance racing vehicles to make cheap, fuel-efficient vehicles for ordinary people.

EARLY YEARS

Soichiro Honda was born in the small village of Komyo, Japan, in 1906. Honda's family was very poor, and five of his eight brothers and sisters died during childhood.

Honda was an indifferent student who constantly got into trouble. He became fascinated at an early age by the foreign cars that occasionally passed through Komyo. When he was a teenager he saw a help-wanted ad in a magazine for an automobile-repair shop in Tokyo—Art Shokai. Honda immediately wrote to the shop and asked for a position as an apprentice. He was hired and went to Tokyo to work in 1922.

Honda's first job at Art Shokai was babysitting for the shop owner's child, but eventually he was trained in car repair, and he also began racing cars. In 1928, Honda founded a branch of Art Shokai in Hamamatsu, a city near his hometown. The Hamamatsu garage did well, and Honda continued to race cars until 1936, when he was seriously injured in a crash.

MANUFACTURING

In 1937 Honda decided that the business prospects for a repair shop were too limited and decided to manufacture piston rings, which are used in engines. He did not know how to make high-quality piston rings, however, and had to return to school to learn. Eventually his business began to prosper.

Japan soon became embroiled in World War II (1939–1945), and Honda's factories were directed by the government to produce goods for the military. During the war, Honda's factory was bombed in a U.S. air raid.

In August 1945, Japan surrendered to the United States. Following Japan's surrender, a larger Japanese firm (now known as Toyota Motor Corp.) asked Honda to continue with his manufacturing business as one of its suppliers. Certain that the American occupying forces would punish large manufacturers like Toyota for having supported the Japanese

Soichiro Honda, photographed just before a hot-air balloon ride in 1980.

war effort, Honda decided to sell his business to Toyota and took the next year off.

AFTER THE WAR

Honda would later say that he was taking a year off to observe postwar Japan and to look for business prospects. The wartime bombings had damaged Japan's roads and destroyed its railways. As a result, food shipments were not reaching cities, and city dwellers had to walk or bike into the countryside to find enough to eat.

Honda took the small army engines left over from the war and attached them to bicycles. The resulting motor-assisted bike made the trips to the countryside easier. Gasoline was scarce; Honda's bikes had such small motors that they did not use much gas, and they could even be run on turpentine oil made from pine trees.

In the fall of 1946, Honda established a small operation that would become the Honda Motor Company. Sales were brisk, and Honda soon ran out of army engines. He designed his own small engine that could either be purchased with a bicycle or clipped onto a bicycle the rider already owned.

TAKEO FUJISAWA

By 1948, Honda's engines were selling well, but his business was still faltering. Honda tried to develop a network of stores that would sell his engines throughout Japan, but he did not keep good records, and he would often deliver engines to retail stores that did not pay for them.

That year, a friend introduced him to Takeo Fujisawa, who had worked as a salesman and knew how to market products and run a business. The two men quickly established an unusual but successful partnership, with Honda retaining nearly-total control over the engineering and design of the company's motors and Fujisawa managing the company's marketing and finance.

Fujisawa quickly began revamping Honda Motor's distribution network, ensuring that the company's products would be widely available in stores. Honda went to work developing the company's first true motorcycle, the Dream. The company moved its headquarters from Hamamatsu to Tokyo in 1950.

The Dream sold relatively well, as did the small clip-on engines that Honda Motor continued to manufacture and improve. However, Honda wanted to do something that would

> Many people dream of success. To me, success can be achieved only through repeated failure and introspection. In fact, success represents the 1 percent of your work that results from the 99 percent that is called failure.
>
> —Soichiro Honda

Sete Gibernau of Spain leads the pack on his Honda motorcycle after the start of the Malaysian Motorcycle Grand Prix in October 2003.

move his company ahead of the roughly two hundred other Japanese companies that were then manufacturing motorcycles.

In 1952 Honda traveled to the United States, which had more advanced manufacturing equipment available than could be found in Japan. There he purchased about $1 million in equipment; at the time, his company was worth less than one-fifth of that amount. The result was a serious financial crisis in 1954. Fujisawa spent several months fending off suppliers to whom money was due and labor unions whose workers faced pay cuts. Fujisawa did manage to pull the company through this difficult period.

THE TOURIST TROPHY

Because of the way the two men divided their responsibilities, Honda did not deal directly with finances. However, he did note that the company's financial difficulties had caused a serious decline in morale among the engineers. He resolved to address this issue in his own way. In 1954, Honda announced that his company would design and build a motorcycle good enough to be entered into the Isle of Man Tourist Trophy race in Great Britain, one of the most competitive motorcycle races in the world.

Honda then traveled to the Isle of Man to examine the motorcycles competing in the race. He was met by a daunting scene: the European and American racing motorcycles were far more advanced than anything he had ever seen in Japan.

Honda felt that he could not back away from his declaration, however, so he returned to Japan and went to work with his engineering team to design a winning motorcycle. Racing motorcycles were classed by the size of their engines, so a designer had to figure out a way to get the most power from an engine without simply making it larger. Honda and his team made a significant adjustment in designing the motorcycle engine to thoroughly burn its gasoline, thus getting the most power possible from the available fuel.

In 1959, Honda Motor made its first entries into the Tourist Trophy, turning in a respectable performance, but not placing first in any races. The same happened the next year. The engineers kept improving the bikes, and in 1961, Honda motorcycles placed first, second, third, fourth, and fifth in two different size classes, drawing the attention of motorcyclists worldwide.

THE STEP-THROUGH

Although Honda had been focusing on racing motorcycles, he had also begun working on another design. In the mid-1950s, Fujisawa had suggested that Honda design a motorcycle not for racers but for ordinary people who needed a reliable, easy-to-use bike.

Motorcycles for export, outside a Honda factory in Japan, 1970.

Honda's Commitment to Excellence: Milestones in Motor Sports

1961	Honda team wins first five places in Isle of Man TT Race in two categories.
1964	Honda makes Formula 1 (F1) debut at German Grand Prix.
1965	Honda earns first F1 win, in Mexico.
1966	Honda claims Constructor's Championship in all classes at World Grand Prix, a world record.
1981	Honda wins 500-cc Motocross World Championship for third year in a row.
1982	Honda XR500R wins Paris-Dakar Rally.
1983	Honda returns to F1 racing after 15-year hiatus.
1984	Honda wins for first time after its F1 return.
1986	Honda finishes first, second, third, fifth, and sixth in Paris-Dakar Rally.
1986–1991	Honda engines take six Formula 1 Grand Prix titles.
1991	Honda posts 10th consecutive win in Isle of Man TT Race.
1992	Honda announces withdrawal from F1 racing.
1993	Honda wins World Solar Challenge, the world's biggest solar car competition, with the Honda Dream.
1994	Honda enters US CART Series.
1996	Honda takes CART Series Manufacturer's Championship.
1996	Honda wins World Solar Challenge 1996.
1999	Honda sweeps first three places in motorcycle TT Formula One race.
2000	Honda begins its third era of F1 racing.
2001	500 World Grand Prix victories in motorcycle racing.
2001	Honda takes sixth consecutive Driver's title in CART Series; fourth Manufacturer's title in CART Series.
2003	Honda enters Indy Car Series.
2004	Honda engines win first through seventh places at Indy 500.
2005–2006	Honda wins Manufacturer's Championship and Driver's Championship.

Motorcycle
Formula 1
CART Series
Indy Car Series
Solar Challenge

Honda designed the SuperCub, often referred to as a step-through motorcycle. The step-through had a very small engine, only 50 cubic centimeters (cc), the same size as Honda's very first clip-on motor for bicycles. The original 50-cc engine had horsepower of only 0.5, but partly because of the technology that the Honda company had developed for its racing motorcycles, the same size engine now had a horsepower of 4.5—as a result, it could power a motorcycle that could be driven fast enough on streets with lots of car traffic.

Although motorcycles had a reputation of being dangerous machines, the step-through was designed to be accessible and safe. Because the engine was small and light, the motorcycle was also light, making it easy to steer. The step-through was reliable and fuel-efficient, so it cost little to maintain. The frame, which was usually brightly colored, dipped down almost to the ground between the seat and the front wheel. That design made the motorcycle easy to mount and dismount, and also made it accessible to riders wearing skirts.

Honda's step-through motorcycle entered the Japanese market in 1958, and it was an immediate hit. Honda Motor later introduced it into foreign markets, including the United States. By 1961, Honda Motor was selling one hundred thousand motorcycles every month, an unprecedented number in the motorcycle industry. The step-through turned Honda Motor into the largest motorcycle manufacturer in the world, a position it has maintained to the present.

DESIGNING CARS

Soichiro Honda, however, was not satisfied with dominating the motorcycle market; he wanted to build cars. Honda Motor produced its first car in 1962. As he did with motorcycles, Honda focused simultaneously on high-speed sports cars and small, efficient passenger cars.

In the early 1970s, however, a new challenge arose: pollution control. In 1970, the U.S. Congress passed a law that would create strict

TIME LINE

1906	1922	1928	1945	1946	1948
Soichiro Honda born in Komyo, Japan.	Honda begins an apprenticeship at Art Shokai in Tokyo.	Honda opens a branch of Art Shokai in Hamamatsu.	Honda sells his business to Toyota.	Honda founds company to build and sell motor-assisted bikes.	Honda meets Takeo Fujisawa.

TIME LINE (continued)

1950	1958	1961	1962	1973	1975	1991
Honda company moves its headquarters to Tokyo.	Honda begins selling the step-through motorcycle in Japan.	Honda motorcycles place first, second, third, fourth, and fifth in two size classes at the Isle of Man races.	Honda produces its first car.	Honda retires.	The Honda Civic compact with CVCC engine goes on sale in the United States.	Honda dies.

emissions controls for automobile exhaust beginning in 1975. California, an important part of the U.S. market, was planning to adopt even more stringent controls.

Honda decided to focus research on making an engine that was very low-polluting. The result was the compound vortex controlled combustion, or CVCC, engine, which in 1972 became the first engine to comply with U.S. emissions standards. Honda began exporting its Civic compact model to the United States with a CVCC engine in 1975.

The CVCC engine produced less pollution because it was designed to burn gasoline more completely than a conventional engine. Such efficiency

Two Honda FCX hydrogen-powered fuel cell cars on display in 2004 in San Francisco, California.

How the CVCC Engine Worked

Honda Motor's compound vortex controlled combustion engine (CVCC) in many ways resembled a conventional car engine. However, it burned gasoline more efficiently and produced less polluting exhaust.

The key to the CVCC's efficiency was the correct mix of fuel and oxygen. When gasoline is burned in a low-oxygen environment, it produces fewer pollutants. If the oxygen in the engine chamber is too low, however, the spark plugs would not ignite the mixture, and the motor would not run. As a result, most automobile engines at the time were designed to contain a large amount of oxygen in the chamber.

The CVCC engine had two chambers. A small chamber near the spark plugs contained a lot of oxygen, to ensure that the fuel-and-air mixture would ignite. Once ignition took place, an opening was created to a larger chamber containing an oxygen-poor mixture, which would then be ignited by the already burning mixture from the first chamber. The result was an engine that combined the reliability of a high-oxygen mixture of gas with the cleanliness of a low-oxygen mixture of gas.

in combustion contributed to greater miles per gallon—the Civic could travel 44 miles (71 km) on a gallon of gasoline, a level of fuel efficiency that was far superior to anything offered on the American market at the time. Gasoline prices were high in the 1970s, and American consumers turned to the Civic in droves. Previously, U.S. car manufacturers had dominated the domestic market; Honda Motor's innovation would help Japanese car manufacturers to become well-established, popular brands in the United States.

Honda retired from Honda Motor along with Fujisawa in 1973, although he continued to advise the company and to promote new technology. In 1991, he died of liver failure in Tokyo at the age of 84.

AFTER HONDA

Honda Motor remains a major car manufacturer and the world's largest motorcycle manufacturer; its annual sales for 2005 were $81 billion. Honda Motor has continued to market more efficient automobiles, including gas-electric hybrids.

Soichiro Honda's impact was broader than simply founding a successful corporation. Honda altered many of the conventions of the motorcycle and automobile industry. He dramatically expanded the market for motorcycles by designing cycles that novices could easily ride. With the CVCC engine, he helped to prove that environment-friendly features, with fuel efficiency, offered a potent marketing point for automobile manufacturers.

—Mary Sisson

Further Reading

Books
Ikeda, Masajiro, ed. *Soichiro Honda: The Endless Racer*. Tokyo: Japan International Cultural Exchange Foundation, 1993.
Sakiya, Tetsuo. *Honda Motor: The Men, the Management, the Machines*. Tokyo: Kodansha International, 1982.
Shook, Robert L. *Honda: An American Success Story*. New York: Prentice Hall, 1988.

Web site
Honda History
    Honda Motor's online exhibition on its history and founder.
    http://world.honda.com/history/index.html

See also: Benz, Karl; Daimler, Gottlieb; Michaux, Pierre; Transportation.

ROBERT HOOKE

Inventor of various scientific
instruments

1635–1703

Although he is credited with many "firsts" in several fields of scientific inquiry, Robert Hooke has remained in relative obscurity for many years. In comparison with that of his contemporary Isaac Newton (1643–1727), Hooke's reputation lapsed considerably in the decades following his death—partly as a result of the rivalry that had grown between Hooke and Newton. The two men worked on many of the same problems and inventions at the same time, and sometimes controversies arose about who had been the first to formulate a particular idea. Nevertheless, Hooke is now appreciated as one of the greatest inventors and scientists of his age.

EARLY YEARS

Robert Hooke was born in Freshwater, on the Isle of Wight, England, on July 18, 1635. His father, the parish curate Reverend John Hooke, intended his son to follow him in a career in the church. However, the young Hooke suffered ill health for much of his childhood. Frequent headaches plagued him, making reading difficult. As a result, John Hooke believed that his son would not live to adulthood and abandoned plans for Robert's formal education.

Left to his own devices, Robert Hooke began to make mechanical toys while developing his observational and mechanical skills. From his own study of his surroundings on the island, Hooke began to form his belief that nature operated as a complex machine. He set about trying to understand as many aspects of this vast machine as possible.

According to the author John Aubrey (1626–1697), Hooke had developed a strong interest in painting when the painter John Hoskyns came to Freshwater to draw. Not yet in his teens, Hooke decided he would learn to draw and paint, and he started to sketch. In 1648, upon the death of his father, Hooke went to London with his inheritance of £100 to study with the celebrated painter Peter Lely (1618–1680). The money was meant for Lely, but Hooke soon determined that he could pursue his goals without the help of Lely and so withdrew from Lely's tutelage. The 13-year-old Hooke decided instead to enter Westminster School.

At Westminster, Hooke studied Greek and Latin, learned to play the organ, and, according to Aubrey, "in one week's time made himself master of the first six books of Euclid"—a reference to the writings of the great Greek mathematician. Hooke's talents for geometry and mechanics advanced rapidly at Westminster, and in 1653 Hooke entered Christ College of Oxford University. There he studied

Robert Hooke's youthful interest in drawing was put to good use in his later scientific work, as this illustration of a fly in Hooke's Micrographia *(1665) shows.*

with many prominent scientists of his time, including the chemist Robert Boyle (1627–1691), with whom Hooke worked especially closely.

FIRST SCIENTIFIC EXPERIMENTATION AND INVENTIONS

Beginning in 1655 Hooke was employed to assist Boyle in his work. He would remain with Boyle until 1662; during this period Hooke made the first in a long line of inventions. Boyle, noticing Hooke's talent for designing and modifying various scientific instruments, asked his assistant to construct an air pump for him to use in determining the laws of gases. The air pump Hooke made was a vast improvement over others then in use and is effectively the same design in use for manual air pumps today.

Also while working under Boyle, Hooke formulated what would become known as Hooke's Law. This law states that an elastic material, such as a spring, will stretch to a degree proportional to the stress exerted on it. This concept, which Hooke used to design a more accurate system of balance springs in watches, among other applications, would remain important in the field of structural engineering into the 21st century.

CURATOR OF EXPERIMENTS

As an indirect result of Boyle's patronage, Hooke was elected to the prestigious post of curator of experiments at the newly founded Royal Society in 1662. In this capacity, Hooke was responsible for supervising all experiments performed at the society's meetings, which featured many prominent scientists. Here Hooke's already abundant range of scientific interests found an even broader arena as dozens of others brought their interests and research to the society's halls.

A reconstruction of Hooke's microscope from around 1665.

While Hooke was curator of experiments, his mechanical skills led to his inventing dozens of new devices, as well as improving the inventions of others. Prominent among these was a reflecting microscope, which surpassed anything that had been constructed to date. Other important inventions from this period include a reflecting telescope, an improved barometer (Hooke was perhaps the first person to associate the drop in barometric pressure with the approach of a storm), the universal joint (sometimes called the Hooke joint), and a simple calculating machine. His experiments demonstrated in rough outline the laws of gravity several years before Newton formulated these laws in more polished form, and Hooke proposed a theory ("inverse-square relationship") to express the decrease of gravity as an object reached a greater and greater distance from earth. He also at this time proposed in rudimentary form a wave theory of light, a dynamical theory of heat, and even an evolutionary theory based on his close study of fossils using his new microscope.

PUBLISHES INFLUENTIAL WORK

The most important result of all these far-reaching inquiries was Hooke's *Micrographia*, published in 1665. In this work, Hooke wrote about his findings on a broad range of phenomena, mostly observations he made with his microscope. Accompanying Hooke's text were many fascinating, often quite beautiful, drawings Hooke himself made of the objects he observed through his microscope. Hooke's *Micrographia* effectively began the field of microbiology; in fact, Hooke coined the term *cell* in this work to describe the structures he observed first in cork and then in plants. These minute, honeycomb-like structures reminded him of monks' quarters (*cellula* in Latin). Hooke is also credited with starting the science of crystals; he was the first to explain the complex structure of crystals, including that of snowflakes, in *Micrographia*.

In 1665, Hooke became professor of geometry at Gresham College in London, a position he held for 30 years. However, the following year a fire devastated much of the city. The Great Fire of 1666 left many of the city's buildings and neighborhoods in ruins. Hooke was named surveyor of the city of London after the fire

In Hooke's Micrographia, Plate XI is entitled "The Texture of Cork"; it shows the honeycomb structures that Hooke called cells.

The Great Fire of 1666

The Great Fire of 1666 caused enormous damage to the city of London. Although relatively few people died in the fire (figures range from 5 to 20), buildings across more than four hundred acres were destroyed. The fire burned for four days, and when it was over, the government was faced with a huge rebuilding project.

Robert Hooke was named surveyor of the city of London and put in charge of laying out the overall plan for reconstruction. Christopher Wren (1632–1723), one of England's greatest architects, was responsible for designing specific buildings. As assistant to Wren, Hooke also participated in the design of a number of important buildings, including the Monument to the Fire, which stands on Fish Street Hill, near where the fire started. Because the two men worked very closely together, determining which of them can be credited with specific concepts or work remains difficult.

Hooke was specifically charged with laying out the street plan for the new London. He proposed a design that included a grid pattern featuring wide boulevards with smaller arteries coming off them. The many property disputes that arose after the damage, however, made Hooke's plan impossible to implement. In fact, he often had to serve as arbitrator between people who were arguing over property lines. London streets were, as a result, rebuilt according to the original medieval design. Many today blame the congestion in London on the property disputes that followed the Great Fire and prevented Hooke's plan from being adopted.

Undated painting of the Great Fire of 1666.

and became chief assistant to the great architect Christopher Wren (1632–1723) in various rebuilding projects. Although his expertise was primarily in mechanical and engineering matters, Hooke also proved himself to be a skilled designer and architect. Among his designs were the Bethlem Royal Hospital (also called "Bedlam," one of the first psychiatric

TIME LINE

1635	1648	1655	1662	1665	1666	1703
Robert Hooke born on the Isle of Wight.	Hooke travels to London to study painting.	Hooke goes to work assisting chemist Robert Boyle.	Hooke becomes curator of experiments of the Royal Society.	Hooke publishes *Micrographia*.	Hooke becomes surveyor of the city of London after the Great Fire.	Hooke dies.

hospitals) as well as the Royal College of Physicians. He also designed, with Wren, the Monument to the Great Fire, which stands near the north end of London Bridge, not far from where the fire began.

Robert Hooke remained active in London until his death in 1703. He delivered numerous addresses to the Royal Society on a wide range of topics, including a discourse on the effects of earthquakes. He also published, among many other papers, "An Attempt to Prove the Motion of the Earth by Observations." Among his inventions during his later years were a type of odometer; an "otocousticon," which was a prototype of the hearing aid; and an underwater telescope. Robert Hooke died in London on March 3, 1703.

—Paul Schellinger

Further Reading

Books
Bennett, Jim, Michael Cooper, Michael Hunter, and Lisa Jardine. *London's Leonardo: The Life and Work of Robert Hooke*. Oxford and New York: Oxford University Press, 2003.

Jardine, Lisa. *The Curious Life of Robert Hooke: The Man Who Measured London*. New York: HarperCollins, 2004.

Web site
Robert Hooke Organization
http://www.roberthooke.org.uk/
Biography of Hooke and material from his writings.

See also: Newton, Isaac; Optics and Vision; Science, Technology, and Mathematics.

GRACE MURRAY HOPPER

Inventor of the computer
language compiler
1906–1992

Only a few decades ago, a person needed a degree in mathematics to operate a computer. In the 1950s a brilliant young American, Grace Murray Hopper, pioneered a way of programming computers (giving them lists of things to do) that was easier to understand. Her invention helped to revolutionize computing for businesspeople, academics, and the military—and earned her the nickname "Amazing Grace," mother of computing.

EARLY YEARS

Hopper was born Grace Brewster Murray on December 9, 1906, and soon gained a reputation as a curious, scientifically minded child. By age seven, she was taking alarm clocks apart to see how they worked. Her father, Walter Fletcher Murray, was successful as an insurance broker despite having had both legs amputated. He passed on a spirit of determination to his three children; the oldest, Grace, learned from him that she could achieve whatever she set her mind to.

After graduating from Vassar College in 1928, Grace Murray went to Yale and earned her master's degree in mathematics in 1930. That year she married Vincent Hopper, an English teacher at New York School of Commerce, changing her name to Grace Murray Hopper. In 1934, she became the first Yale woman to earn a PhD in mathematics. Dr. Hopper returned to Vassar as a math teacher, becoming an associate professor in 1941.

COMPUTER PIONEER

When the United States entered World War II, Hopper was determined to contribute to the war effort. In December 1943, she was accepted by the U.S. Naval Reserve and assigned to Harvard University. There, she joined a small team that used math to solve problems such as how to protect ships from mines. Her boss, computer pioneer Howard Aiken (1900–1973), greeted her with the words: "Where the hell have you been?" He set her to work on one of the world's earliest electro-mechanical computers, the Harvard Mark I. It was at this time that she wrote some of the first programs for the 51-foot (15.5 m) machine. Hopper also wrote its 500-page operating manual.

Grace Murray Hopper, photographed around 1961.

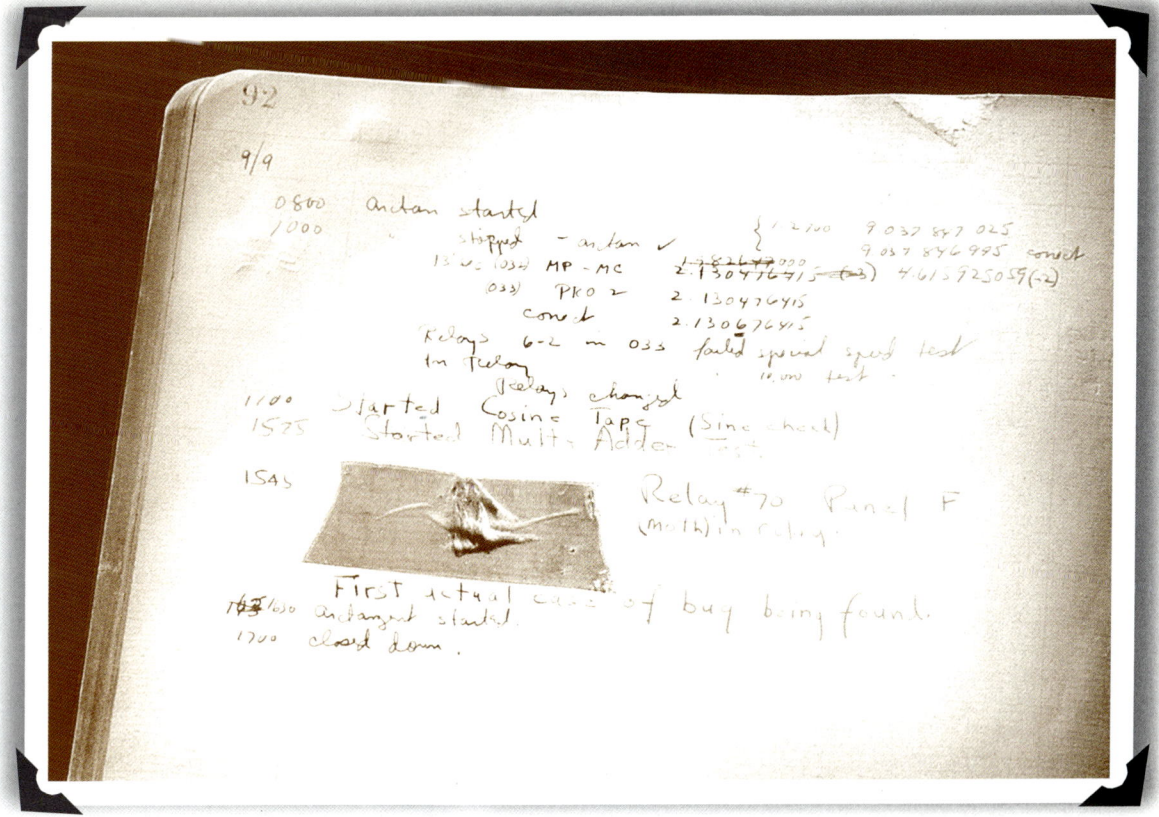

A page from Hopper's logbook, which includes the actual moth Hopper removed from a computer, helping to popularize the term bug for a computer problem.

World War II brought tragedy to millions, including Hopper, whose husband was killed. After the war ended, she stayed on at Harvard to develop the improved Harvard Mark II and Mark III computers. During this time, she helped popularize the term *bug*, meaning a programming error. Moths often flew inside the huge computers and stopped them from working. Removing the bug—or debugging—got the machine running again.

At this time, computers were used largely by universities and the military. Many people believed that no real market for business computers existed. Hopper thought differently. In 1949, she took a gamble with her promising career and joined a small company, Eckert-Mauchly Computer Corporation, founded by computer scientists John Mauchly (1907–1980) and J. Presper Eckert (1919–1995). There she helped to design the UNIVAC-1, the world's first large commercial electronic computer.

COMPUTER LANGUAGES

Complexity was one reason most people did not use early computers. Hopper realized computers would be employed in many more fields if they were easier to program and use, so she developed a compiler. She envisaged that people would write a program (the sequence of instruc-

TIME LINE

1906	1934	1943	1949
Hopper born in New York City.	Hopper receives a PhD from Yale University.	Hopper helps design the Harvard Mark I.	Hopper helps design the UNIVAC-1.

tions for a computer to carry out) in meaningful symbols or English words. Her compiler would translate this program into raw, numeric instructions (machine code) that the computer could process. So the compiler was a kind of translator that helped computers and people to "understand" one another. Hopper worked for three years to convince her colleagues about the feasibility of her idea; she finally published the concept in 1952.

During the 1950s, Hopper took her ideas for compilers a step farther and helped to develop the first major computer programming language for business, COBOL (Common Business Oriented Language). This enabled people to write computer programs for everyday business problems like preparing payrolls or billing clients using strings of simple English commands. Hopper joked that she helped to develop COBOL "because I couldn't balance my checkbook."

REAR ADMIRAL HOPPER

From the end of World War II, Hopper remained in the Naval Reserve and combined her military work with her business career. In December 1966, she was obliged to retire from the navy after reaching the age of 60. The navy soon realized it could not manage without her, so she was summoned back "just for six months" to sort out its computing problems. Almost 20 years later, she was still there. At age 79, she was the oldest serving officer and was given a grand retirement ceremony on the deck of the USS *Constitution*. By this time, she had been promoted to rear admiral—one of the most senior positions in the navy.

Although technically retired, Hopper immediately began a new career as a senior consultant to the computer maker Digital Equipment Corporation, touring the world to give lectures and spread her ideas. She drew a large, admir-

> The most damaging phrase in the language is: "We've always done it this way."
>
> —Grace Murray Hopper

TIME LINE

1952	1950s	1985	1992
Hopper develops and publishes her ideas for the first compiler.	Hopper develops COBOL.	Hopper retires from the navy as a rear admiral.	Hopper dies.

ing audience wherever she went and continued this work until about eighteen months before she died on January 1, 1992.

Rear admiral was only one of many honors "Amazing Grace" Murray Hopper received. Ironically, her first major award was to be named "Man of the Year" by the Data Processing Management Association in 1969. The navy named a ship for her—the USS *Hopper*—in 1996. Grace Hopper is remembered as an original thinker: she had a clock in her office

Hopper in her office in 1984.

whose hands spun counterclockwise to challenge people to look at problems in new ways. She was a visionary who realized that computers could benefit ordinary people if they were easier to use. She was also an inspiration to many: her achievements proved that women can be just as successful as men in scientific and technical fields, as well as in the military.

—Chris Woodford

Further Reading

Murphy, Patricia. *Grace Hopper: Computer Whiz*. Hillside, NJ: Enslow, 2004.
Whitelaw, Nancy. *Grace Hopper: Programming Pioneer*. San Francisco: Freeman, 1995.

See also: Computers; History of Invention; Science, Technology, and Mathematics.

HOUSEHOLD INVENTIONS

That homes are places of comfort and convenience is largely attributable to inventors and their many practical inventions. Everything—from the doorbell that announces visitors, to the mower that makes the lawn neat—was invented by someone. Many devices were developed to speed up chores, while some were developed purely for entertainment. A few make homes safer, while others make them cleaner and more hygienic. Modern homes are packed with so many useful gadgets and appliances that they are arguably living museums of invention.

Most household inventions are relatively recent, dating from the late 19th or early 20th century. The reason is simple: many of them depend upon electricity, which became popular only after the prolific inventor Thomas Edison (1847–1931) developed the first power plants in the 1880s. Edison did not know that modern homes would become so reliant on electricity; his objective was simply to sell the idea of electric-powered lighting. Intentional or not, the development of electric power was a milestone in the history of the household that made possible all kinds of labor-saving appliances, freeing people to spend more time with their families or on leisure pursuits.

HEATING AND COOLING

Fire was one of the first human inventions, dating back at least a million years.

The underground tunnels of an ancient hypocaust, found at an archaeological site in Sbeitla, Tunisia.

No one knows who invented it or how, but its uses are clear: to provide warmth, safety, and a means of cooking food. Fire, in one form or another, has maintained a steady presence in the home ever since. In Roman times (ca. 27 BCE—395 CE), some homes were heated by hot air produced by an underground fire and channeled between the walls and floors using an ingenious network of brick-lined ducts. Called a hypocaust, which means "heated from below," it was the earliest example of central heating.

Until the mid-18th century, most people heated their homes using large open fireplaces. In 1740, the famous American journalist, politician, and inventor Benjamin Franklin (1706–1790) developed what became known as the Franklin stove. This was a large, freestanding iron basket into which wood or other fuel could be loaded. Air could be drawn through the stove freely, thus burning the fuel more effectively and giving off less smoke and more heat. Walter Hunt (1796–1859) was another American inventor who developed a household stove, although he is best known for inventing the safety pin.

English physicist Benjamin Thompson (1753–1814), one of Benjamin Franklin's contemporaries, gained notoriety for his scientific studies of heat. Thompson was one of the first to show that heat is a form of energy. He made his scientific discoveries practical with a number of inventions that improved home heating and cooking, such as a water boiler, a kitchen range, and an improved fireplace.

Inventions that cool homes were also developed through a mixture of trial and error and scientific discovery. One of the simplest home-cooling devices, the electric desk fan, was invented in the 1880s by American scientist Schuyler Wheeler, just four years after Edison's first power plant opened in Manhattan. Most fans move the air around—they do not actually remove any heat—and so the cooling they provide is sometimes an illusion. Air-conditioning, a much-improved way of cooling a building using scientific principles, was pio-

Inventor Willis Carrier improved the traditional fan when he discovered how to cool air, rather than move it around.

neered in the first half of the 20th century by American engineer Willis Haviland Carrier (1876–1950). His systems used special cooling fluids (coolants) to soak up the heat in a room and transfer it outside.

PLAYING AND WORKING AT HOME

Inventors have enabled us to make our homes places in which to entertain and relax. One invention in particular, the electronic transistor (a tiny electrical switch and signal booster), has played an enormous part in modernizing homes. It was invented in 1947 by John Bardeen (1908–1991), Walter Brattain (1902–1987), and William Shockley (1910–1989), and one of its first applications was in portable transistor radios. These became popular through the efforts of Japanese industrialist Akio Morita (1921–1999), founder of the Sony Corporation. Later, transistors found their way into such devices as video recorders, invented in the 1950s by Charles Ginsburg (1920–1992), and the popular Sony Walkman, a portable tape player also conceived with help from Morita.

Before home computers really took off in the 1980s, many people wrote letters at home with typewriters, personal printing machines with a keyboard similar to that of a modern computer. Typewriters were developed in the 19th century by Christopher Latham Sholes (1819–1890). When letter keys were pressed on a typewriter, they immediately imprinted the letter onto the paper. Typists had to remove their mistakes from the paper with a special eraser or paint over them with a white correcting fluid (brand name Liquid Paper) invented in the 1950s by Bette Nesmith Graham (1922–1980).

Since the invention of the Internet, homes are increasingly places where people do their work. Most American homes now have a personal computer—something unthinkable as recently as the early 1970s. Home computers became possible as a result of the efforts of many people, but one stands out. During the 1960s, American computer scientist Douglas Engelbart (1925–) developed a whole raft of inventions that transformed computers from forbidding, mathematical monsters into "user-friendly" personal helpers. Among his ideas were the computer mouse and on-screen word processing, two inventions employed by modern computer users all the time.

IN THE KITCHEN

The kitchen is the hub of many homes, a place where people cook, eat, and coordinate domestic chores. Food arrives in the kitchen from grocery stores, often still carried in the flat-bottomed paper bags that became popular after Margaret Knight (1838–1914) invented a machine for making them in the 1860s. Once in the kitchen, food needs to be kept fresh. Most kitchens have airtight plastic containers for preserving a variety of foods. These were developed in the mid-20th century by American inventor Earl Tupper (1907–1983). Cellophane (a thin, clear plastic film) is another way of keeping food fresh. It was invented accidentally by Swiss scientist Jacques Brandenberger (1872–1954), who was

Who Invented These Common Household Items and When?

Computer mouse
Douglas Engelbart
1971

Air conditioner
Willis Haviland Carrier
1906

Flush toilet
Alexander Cummings
1775

Tupperware
Earl Tupper
1947

Color television
Peter Goldmark
1940

Flat-bottomed paper bags
Margaret Knight
1870

Dishwasher
Josephine Cochran
1886

trying to devise a chemical coating that would keep tablecloths clean. The plastic he invented turned out to be more useful for preserving food by keeping the air away from it.

Most modern kitchens have refrigerators and freezers to store fresh and frozen food. Refrigerators use coolants that circulate to remove excess heat, thus preserving the food inside. The coolants, made from chemicals known as chlorofluorocarbons (CFCs), were developed in 1928 by American chemist Thomas Midgley (1889–1944) and first used in refrigerators made by the Frigidaire Company. However, CFCs were largely phased out in the 1990s after scientists found they caused a hole in the ozone layer, an important part of earth's atmosphere.

Home freezing is possible not just because of the invention of refrigerators

The Kitchen Debate

After World War II ended in 1945, the world's two superpowers, the United States and the Soviet Union (now Russia and its neighboring republics), became locked in a hostile decades-long period of history known as the Cold War. Although fighting never broke out during that period, relations between the powers remained tense. Sometimes the rivalry spurred invention. When the Soviet Union launched its Sputnik space satellite in 1957, the United States was determined not to be outdone. Four years later, President John F. Kennedy announced his bold intention to put a man on the moon.

At other times, the Cold War led to almost comical exchanges. Such was the case when Vice President Richard M. Nixon visited the Soviet capital, Moscow, in July 1959. During his stay, Nixon attended a U.S. trade fair with the Soviet premier Nikita Khrushchev. As they moved past a model home where the latest household appliances were on display, they began a lively debate about which of the world's two major political systems was better. Was it capitalism, as practiced by the United States; or communism, favored by the Soviet Union?

The "kitchen debate," as this came to be known, centered on household appliances. First, Nixon mentioned the invention of color television and insisted that his country had taken the lead with this technology. Khrushchev quickly replied that the Soviets had already developed the same idea. Next, Nixon stopped Khrushchev in a model of a kitchen with all the latest appliances. When Nixon proudly pointed out the latest American washing machine, Khrushchev quickly countered: "We have such things." Then Nixon pointed out a television that could be used to watch what was happening in different parts of the home, but Khrushchev was unimpressed: "Don't you have a machine that puts food into the mouth and pushes it down? Many things you've shown us are interesting but they are not needed in life. They have no useful purpose. They are merely gadgets." As they walked on, the two leaders continued a strange and lively war of words, intermingling mundane thoughts about domestic technology with the stern rhetoric of war and peace.

Soviet premier Nikita Khrushchev (second from left), argues with U.S. Vice President Richard Nixon (to his left, in black), beside the model kitchen display at a trade fair in 1959.

HOUSEHOLD INVENTIONS 831

and coolants, but also because American inventor Clarence Birdseye (1886–1956) found a way of "flash freezing" food (cooling it very quickly) in a factory setting. Before frozen food became popular, people had to rely on other methods of preservation. One of the most popular was to seal food in cans, an idea invented in 1810 by French chef Nicolas-François Appert (ca. 1750–1841). Various methods of preserving meat, spices, and oils were also developed by African American inventor Lloyd Hall (1894–1971).

Kitchens contain many of a home's electrical appliances besides the refrigerator—and most of them save huge amounts of time and energy. One appliance often stored in the kitchen is a vacuum cleaner, an invention originally developed by a British engineer in 1901. Until the 1980s, virtually every vacuum caught the dust it collected in a cloth or paper bag, which tended to reduce the cleaner's effectiveness as it filled up. In the 1980s another British engineer, James Dyson (1947–), began working on a method of vacuuming without the dust bag so a vacuum cleaner would always maintain maximum suction. Dyson's bagless vacuum was successfully launched in the United States in 2002.

Another popular kitchen appliance, the dishwasher, was invented in 1886 by Josephine Cochran (1839–1913). The first automatic dishwashers were entirely mechanical: plates, glasses, cutlery, and other items to be washed were loaded into metal baskets and blasted with hot soapy water to get them clean. Modern dishwashers, which still use hot water and detergent, are powered by electricity and controlled by a microchip. This miniature electronic brain was developed in the late 1960s by American electrical engineer Marcian Edward (Ted) Hoff (1937–), working for the Intel Corporation.

Microchips have since found their way into all manner of household appliances. One of the most familiar is the microwave oven, developed in the early 1950s by American electrical engineer,

A salesman demonstrates the effectiveness of a General Electric vacuum cleaner in 1955.

Although this bathtub toy is known as a "rubber duckie," it is actually made out of PVC, not rubber.

Percy Spencer (1894–1970). He was experimenting with radar, a type of radio used in ship and airplane navigation, when he found that radar's waves generated enough heat to melt a candy bar in his pocket. Spencer almost immediately began to concentrate on his accidental discovery—turning it into a revolutionary new technology for cooking food in a fraction of the time needed by conventional ovens.

IN THE BATHROOM

The bathroom contains a surprising number of inventions. One of the simplest, the nylon toothbrush, is actually a product of highly sophisticated modern chemistry. Nylon was developed in the 1930s by American chemist Wallace Carothers (1896–1937) and his coworkers. The technology behind nylon, polymerization, involves building large molecules from many smaller ones based on the chemical element carbon. Apart from nylon toothbrushes, polymerization is used to make other modern plastics, including PVC (polyvinylchloride)—from which such objects as bathtub toys are manufactured.

Plastics are also used to make disposable safety razors, which were first developed in the late 19th century by King C. Gillette (1855–1932). Before the invention of the safety razor, most men shaved with "cutthroat" razors, which were like very sharp knives with rounded blades. These traditional razors were dangerous and needed regular sharpening. Gillette had the idea to make a razor with disposable blades that could be thrown out when they became dull.

The flush toilet is a central feature of every modern bathroom and, in its present form, was developed in 1775 by

TIME LINE

27 BCE – 395 CE	1740	1775	Early 1800s	1821
Hypocausts provide central heating in Roman Empire homes.	Benjamin Franklin develops the Franklin stove.	Alexander Cummings develops the flush toilet.	Nicolas-François Appert develops his method of food preservation.	Michael Faraday develops the basic principles of the electric motor.

British inventor Alexander Cummings. Toilet paper, an even older invention, is believed to date from fourteenth-century China. Modern toilet paper has been manufactured in the United States since 1857, when Joseph Gayetty set up a factory specifically for making the product. Around 1880, the British Perforated Paper Company began making toilet paper in convenient, ready-cut squares. A century later, Japanese inventors developed "paperless" toilets. Ultra-hygienic, they are made from metals that resist bacteria and have a toilet and air dryer built into a single unit that entirely takes over the function of toilet paper.

WOMEN IN THE HOME

Toward the end of the 19th century, the arrival of electric power began a new age that helped to liberate women from domestic chores. One invention played an especially important part: the electric motor. An electric motor uses electricity to generate magnetism, which makes the central part of the motor rotate so it can drive a machine. The basic principle of the motor was developed in 1821 by English scientist Michael Faraday (1791–1867) and turned into a more practical invention about ten years later by another Englishman, William Sturgeon (1783–1850). Motors soon made possible electrical appliances such as the vacuum cleaner (1901), food mixer (1904), washing machine (1909), electric dishwasher (1912), and modern refrigerator (1928).

Well into the 20th century, women were the most likely members of their households to do the domestic chores, and, accordingly, they were the chief beneficiaries of household inventions. The corporations making domestic appliances knew this and advertised their products as both saving time and lightening labor. In 1917, for example, the General Electric Company sold its Mazda light as "The Lamp That Lights the Way to Lighter Housework."

As a result of laborsaving devices, women had more time to devote to their family, to leisure interests, or to careers outside the home. Ironically, women and men in the 21st century still spend roughly as much time on the household chores as they did decades ago. One explanation offered by academics is that people now own more clothes and other possessions than they used to, requiring more hours of housework. Cookery is more sophisticated and a wider range of

TIME LINE (continued)

1832	1860s–1870s	Late 1860s	1880s	1886	Late 1800s
William Sturgeon creates the commutator.	Christopher Latham Sholes develops the typewriter.	Margaret Knight invents a machine to make flat-bottomed paper bags.	Thomas Edison develops the first power plants.	Schuyler Wheeler invents the electric desk fan.	King Camp Gillette develops the disposable razor.

An advertisement for the New Empire Wringer from the 1880s attempted to appeal to women.

TIME LINE

1902	1908	1923	1930s	1930s-1950s	1940s
Willis Haviland Carrier invents air-conditioning.	Jacques Brandenberger perfects cellophane.	Clarence Birdseye discovers how to flash-freeze food.	Wallace Carothers invents nylon.	Lloyd Hall develops several food preservation methods.	Earl S. Tupper develops Tupperware.

ingredients is available from stores, leading amateur cooks to become more adventurous. Another explanation is that the automobile has prompted parents to spend more time transporting children, more than making up for the time they save on other chores. Another theory is that advertisements play on people's guilt, making them feel they must spend longer than necessary on chores.

—Chris Woodford

Further Reading

Books

Cowan, Ruth Schwartz. *More Work for Mother: The Ironies of Household Technology from the Open Hearth to the Microwave*. New York: Basic Books, 1985.

Jones, Charlotte. *Mistakes That Worked*. New York: Doubleday, 1994.

Landau, Elaine. *The History of Everyday Life*. Breckenridge, CO: Twenty-First Century, 2005.

Parker, Steve. *Eyewitness: Electricity*. New York: Dorling Kindersley, 2005.

Web sites

Inventors from About.com
   A Web site that describes many familiar household inventions.
   http://inventors.about.com/

Socket to Me! How Electricity Came to Be
   The story of electricity and the familiar household appliances it spawned.
   http://www.ieee-virtual-museum.org/

See also: Appert, Nicolas-François; Bardeen, John, Walter Brattain, and William Shockley; Birdseye, Clarence; Brandenberger, Jacques; Carothers, Wallace; Carrier, Willis; Cochran, Josephine; Dyson, James; Edison, Thomas; Engelbart, Douglas; Faraday, Michael; Franklin, Benjamin; Gillette, King C.; Ginsburg, Charles; Hall, Lloyd A.; Hunt, Walter; Knight, Margaret; Morita, Akio; Sholes, Christopher Latham; Sturgeon, William; Tupper, Earl.

TIME LINE (continued)

1947	1950s	1957	1960s	1979	2002
Bardeen, Brattain, and Shockley invent the transistor.	Charles Ginsburg develops the video recorder.	Bette Nesmith Graham invents Liquid Paper.	Douglas Engelbart develops user-friendly computer devices.	Akio Morita develops the Sony Walkman.	James Dyson introduces his bagless vacuum to the U.S. marketplace.

ELIAS HOWE

Inventor of the sewing machine
1819–1867

Most people wear clothes that are stitched together by machine, but this was not always the case. Once, all textiles had to be sewn together slowly and laboriously by hand. American inventor Elias Howe made it possible to manufacture clothes rapidly and inexpensively when he developed his sewing machine in 1845. Like many other shrewd inventors, Howe developed a machine to fill a gap in the market and eventually made a considerable amount of money. Many years and much struggle preceded his success, however.

EARLY YEARS

Elias Howe died a millionaire, but he was born into a life of poverty on July 9, 1819. He spent most of his childhood helping his father, a farmer and miller in Spencer, Massachusetts. The farm needed as many workers as possible, so Elias attended school only during the dark winter months. Consequently, he had little formal schooling and did not continue his education beyond grade school. In 1835, at age 16, he became an apprentice at a local textile mill, earning about fifty cents a day. He worked there for two years until the financial panic of 1837 threw many people out of work throughout the United States. Desperate to earn a living, Howe moved to the city of Lowell, Massachusetts, and found work in a machine shop owned by a scientific-instrument maker.

Howe married in 1841, and he and his wife had three children. Howe was sickly, and the strain of supporting a family impaired his health, so his wife started to sew to earn money. It is not known how her husband came upon the idea of making a machine that could automate this job. Some historians believe he overheard a chance remark in the machine shop to the effect that whoever could invent a sewing machine would become rich. Others think he was inspired by watching his wife sew at home. Possibly it was a combination of these.

SEWING BEFORE HOWE

People have been stitching textiles since the late Stone Age, for around twenty thousand years, although the modern process of sewing with metal needles is much more recent—only about five to six hundred years old. Thomas Saint (dates unknown), a British cabinetmaker, became the first person to automate the procedure in 1790. His machine was designed for sewing tough materials such as canvas and leather, which are too thick to sew by hand. First, an awl (punching tool) made a hole in the material; then a rod pushed

An oil painting (artist unidentified) of Elias Howe from around 1850.

Illustrations from Thimmonier's U.S. patent application; although the machine was the only early sewing machine that worked well, by the time the patent was granted in 1850, Thimmonier's invention was already obsolete.

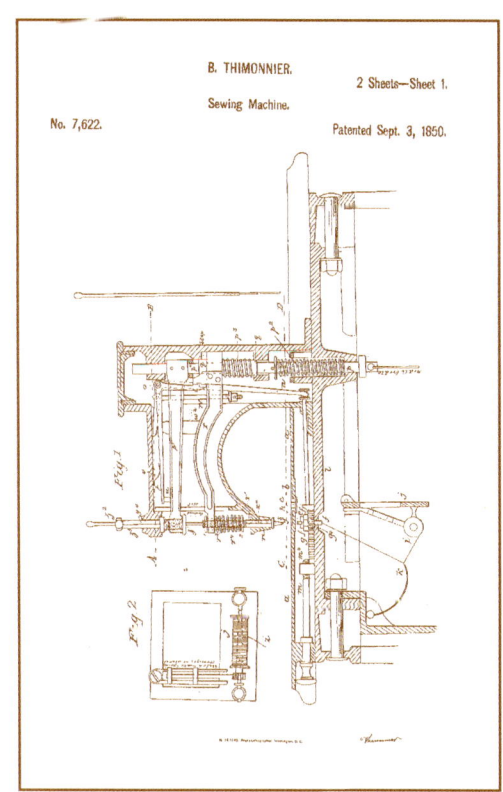

a thread through the hole to make a stitch. This was repeated over and over to fasten two pieces of thick material together. Saint patented his idea, but it is unclear whether he ever made a working machine, and he never put the device into production.

Numerous inventors applied themselves to the problem in the 19th century. A German tailor, Balthasar Krems (1760–1813), made an unsuccessful machine for sewing caps in 1810. An Austrian tailor, Josef Madersperger (1768–1850), was granted a patent for a sewing mechanism four years later, though the various machines he developed from it also failed to work effectively.

The most promising of the early European sewing machines—indeed, the only one that worked reasonably well—was developed in France in 1829 by tailor Barthélemy Thimonnier (1793–1859) and patented the following year. His machine used a metal needle that pushed a thread in and out of the cloth. The French government used 80 of these devices for making soldiers' uniforms, but the machine's success soon caused trouble. Fearing unemployment, a group of French tailors went on a rampage and destroyed Thimonnier's machines, forcing the inventor to flee to England, where he later died in poverty.

Several years later Walter Hunt (1796–1859), the inventor of the safety pin, built the first successful sewing machine in the United States. He realized the machine would put many people out of work and abandoned the idea without even patenting it. Perhaps he had heard Thimonnier's story and worried that a similar fate would befall him.

SUCCESS AND FAILURE

In his mid-twenties, Elias Howe started to devote every spare moment to building a successful sewing machine. For the next four or five years, working several hours each evening after returning home from his job, Howe tried various needle designs, some with points at both ends and the thread held in the middle, and many different mechanisms. Soon, his obsession with the sewing machine had taken over his life. A combination of poor health and his desire to finish the invention made him give up his job; this decision led to financial disaster. He moved his wife and children into his father's home to save money. Then, fortunately, he managed to interest an old school friend, George Fisher, in the invention. Fisher lent him money and saved the Howes from ruin.

Finally, in 1845, Howe developed a successful sewing machine using a clever idea: he found he could mechanize the process of stitching materials together if he used not one thread but two. His sewing machine had a needle with a thread running through it on one side of the two fabrics being joined. On the other side of the fabrics was a shuttle that moved back and forth carrying a second thread. To make a stitch, the needle pushed through the fabrics, carrying the first thread with it. Once it had pierced the material, it made a loop with the thread. At this point, the shuttle slipped out and passed the second thread through the loop. The

An illustration of Howe's sewing machine from 1846.

needle then pulled back through the material, making a tight lockstitch. Turning a hand crank repeated this process over and over to fasten the fabrics securely together. Howe patented this idea in 1846.

Howe's machine was both a fantastic success and a dismal failure. It could sew 250 stitches a minute—five times faster than a skilled seamstress. Yet it cost $300 (an enormous amount in 1846) and no one could afford to buy it. Howe built four of the machines and failed to sell a single one. To add to his woes, his workshop burned to the ground, and he was desperately short of money. He worked on the railroads for a time as an engineer, and then his health failed again.

In 1847, he gave up on the United States and decided to try his luck in England. Howe moved his family across the ocean but fared no better. He sold the British rights to the sewing machine for a small sum and adapted his machine so it could be used to make corsets, umbrellas, and valises. He failed to make a living, however, and sent his wife and children back to the United States. After two years, he had no money at all. He decided to return to America also and had to work his way back by getting a job as a sailor.

DEFENDING HIS PATENT

If Howe expected his life to get better in the United States, he was soon disappointed. When he returned, he found that his wife was dying. In addition, others had started copying his sewing machine and selling it in large quantities. His invention was now a great success—but it was making money for other people. One of them was Isaac Merrit Singer (1811–1875), a shrewd businessman who was able to market his products so that people wanted to buy them (see box, Isaac Merrit Singer). Singer patented his own sewing machine, similar to Howe's, in 1851. Some parts of Singer's machine were completely original, but it used Howe's method of lockstitching the cloth together with two different threads—and that was the most important aspect. As Howe saw it, this was a flagrant violation of his patent.

Singer's success clearly demonstrated that a great deal of money could be made from sewing machines. Howe's father mortgaged the family farm to help raise money for a lawsuit to defend Howe's patent. In 1854, Howe sued Singer for patent infringement. Singer countered that Howe's invention was not itself original. He pointed out that Howe's ideas were very similar to those of Walter Hunt. Crucially, though, Hunt had never patented his own invention. That meant that Howe was effectively the inventor of the sewing machine as far as the law was concerned—and Singer lost the case. Ironically, Singer would have won if Hunt had patented his invention, because Howe's machine would then have been ruled unoriginal.

Isaac Merrit Singer

Elias Howe invented the original stitching mechanism at the heart of the sewing machine, but Isaac Singer was an innovator of a different kind. Singer was born in Pittstown, New York in 1811. He had initially dreamed of being an actor, and he used his showman's talent to great effect when he started making sewing machines, originally just to earn a living.

Singer was not just a marketing whiz, however; he introduced many technical improvements into Howe's basic sewing machine. Singer's version drove the needle up and down instead of sideways, from front to back, as Howe's did. As a result, he could build his machines into a table with the material flowing horizontally across it, so they were easier to use. He made his machines foot-operated instead of hand-cranked, so people could use both hands to steer the material. One of his best ideas was the presser foot, a small piece of metal that presses the materials being sewn closely together, so they can be fed smoothly through the machine.

Mechanical improvements would not, by themselves, have been enough to make the sewing machine a winning invention. Howe's machines had failed to sell; Singer succeeded by making his machines attractive and affordable. Crucially, he allowed people to pay the $75 to $100 asking price (one-fourth the amount Howe had originally charged, but still a large sum) in a number of payments spread over a longer period; he was the first person to let people buy goods on the installment plan. In cities, he sold his machines in showrooms; in rural areas, he used traveling salesmen on commission.

Losing a patent battle with Elias Howe in 1854 proved to be no setback. The following year, I. M. Singer and Company started to sell machines abroad and quickly became the world's biggest manufacturer of sewing machines. Although Singer died in 1875, his company continued to innovate and produced the first electric-powered sewing machine in 1885. The company declined during the 20th century and it was sold several times, but the Singer brand name survives and remains a familiar sight on sewing machines.

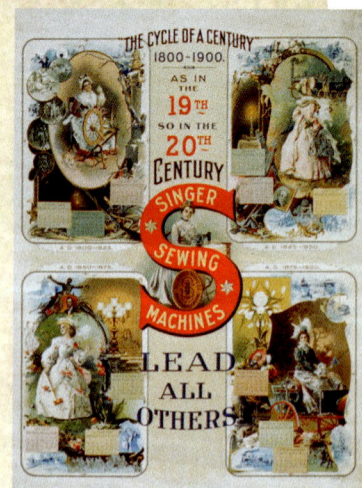

Undated advertisement for Singer sewing machines, probably from around 1900.

WINNING INVENTOR

Following his successful lawsuit, Elias Howe was allowed to earn royalties from Singer and other sewing-machine manufacturers who used his ideas. He earned five dollars on each machine sold in the United States and a dollar on each machine sold abroad. Before the case, he was earning about $300 a year; after his victory, his income rose to around $200,000. Singer lost little through the deal: his marketing genius meant that he sold huge numbers of machines and earned a fortune himself.

After the U.S. Civil War broke out in 1861, Howe enlisted as a private in the 17th Connecticut Volunteers. When poor health cut short his military career, he supported the Union Army in other ways, paying for military equipment and expenses. Between 1854 and 1867, when his patent expired, he earned around two million dollars in royalties. Yet ill health and struggle finally took their toll, and he died the same year, on October 3, in Brooklyn, New York, at the age of 48.

IMPACT OF THE SEWING MACHINE

It is often the case with inventions that, once someone has made a breakthrough by solving a crucial problem, others can develop even better devices very rapidly. Isaac Singer made a whole series of innovations, improving the way sewing machines were designed and sold. Others followed suit. Another American inventor, Allen B. Wilson (1824–1888), developed a better mechanism for feeding the material and the rotary bobbin (a small spool) for storing the thread. Machines were soon developed that could sew a variety of different stitches: American farmer James Gibbs (1829–1902) invented a machine that could chain-stitch in 1857; Helen Blanchard (1840–1922) developed the first machine that could sew a zigzag stitch for finishing seams in 1873.

Many people saw that the sewing machine would bring changes to society. Yet the outcome was more positive than many had feared. The

TIME LINE

1819	1835	1837	1845	1854	1861	1867
Elias Howe born in Spencer, Massachusetts.	Howe is apprenticed at a textile mill.	Howe moves to Lowell, Massachusetts, and works in a machine shop.	Howe develops a successful sewing machine.	Howe sues Isaac M. Singer for patent infringement and wins.	Howe enlists as a private during the U.S. Civil War.	Howe dies.

A woman uses a sewing machine in Kibera, Kenya, in 2004.

sewing machine was the first great domestic appliance. Families that had sewing machines could provide themselves with a better range of clothes much less expensively. As Elias Howe found, however, early machines were initially more expensive than most households could afford. Isaac Singer made a crucial innovation by selling his machines on the installment plan, so people could pay for their machines a little at a time over a long period; sometimes families, friends, or neighbors joined together to buy machines they could share.

The sewing machine brought social changes beyond the home. The cotton gin that Eli Whitney (1765–1825) invented in the 1790s had enormously expanded the quantity of textiles available. Fifty years later, factory owners used the sewing machine in making this abundance of cloth into clothes in great quantity. Although Elias Howe fought in the Civil War and donated money to the Union, his biggest contribution to the war effort was probably through the use of sewing machines to make military uniforms in standard sizes. After the 1860s, the idea of a person making all his or her own clothes became less and less fashionable as more people bought manufactured garments in just a few fixed sizes. The modern clothing industry was born—thanks, in no small part, to the hard work and perseverance of Elias Howe.

—Chris Woodford

Further Reading

Brandon, Ruth. *Singer and the Sewing Machine: A Capitalist Romance*. New York: Kodansha, 1996.

Carlson, Laurie. *Queen of Inventions: How the Sewing Machine Changed the World*. Brookfield, CT: Millbrook, 2003.

Steele, Philip. *Clothes and Crafts in Victorian Times*. Milwaukee, WI: Gareth Stevens, 2000.

See also: Cloth and Apparel; Hunt, Walter; Whitney, Eli.

WALTER HUNT

Inventor of the safety pin

1796–1859

Today, few people can answer the question, "Who was Walter Hunt?" Yet Walter Hunt was one of the most prolific inventors of the 19th century, credited with dozens of devices, some patented and others not. Although he invented the first sewing machine, almost twenty years before Elias Howe received credit, Hunt is perhaps best known for inventing the safety pin.

EARLY YEARS

As with many other mostly forgotten inventors, very little is known about Hunt's life. He was born near Martinsburg, New York, and lived most of his life in New York City. When and whom he married are unclear, but he certainly had a family. In 1935, Clinton N. Hunt, a great-grandson, self-published a pamphlet that included descriptions of more than two dozen of Hunt's inventions. Hunt was by trade a mechanic. Most accounts suggest that, despite his wealth of inventions, he never became wealthy himself.

SEWN SHUT

Hunt invented the first lockstitch sewing machine around 1834. It was not the first sewing machine, but it was the first machine to utilize lock-stitches, in which two threads interlock at a seam, instead of trying to mimic hand-sewn stitches. Hunt's machine was also the first to utilize a curved, eye-pointed needle—another invention by Hunt.

Hunt never patented his sewing machine. He allegedly lost interest when he thought the machine would cause unemployment by putting seamstresses and tailors out of work. (Such an outcome was unlikely, as his

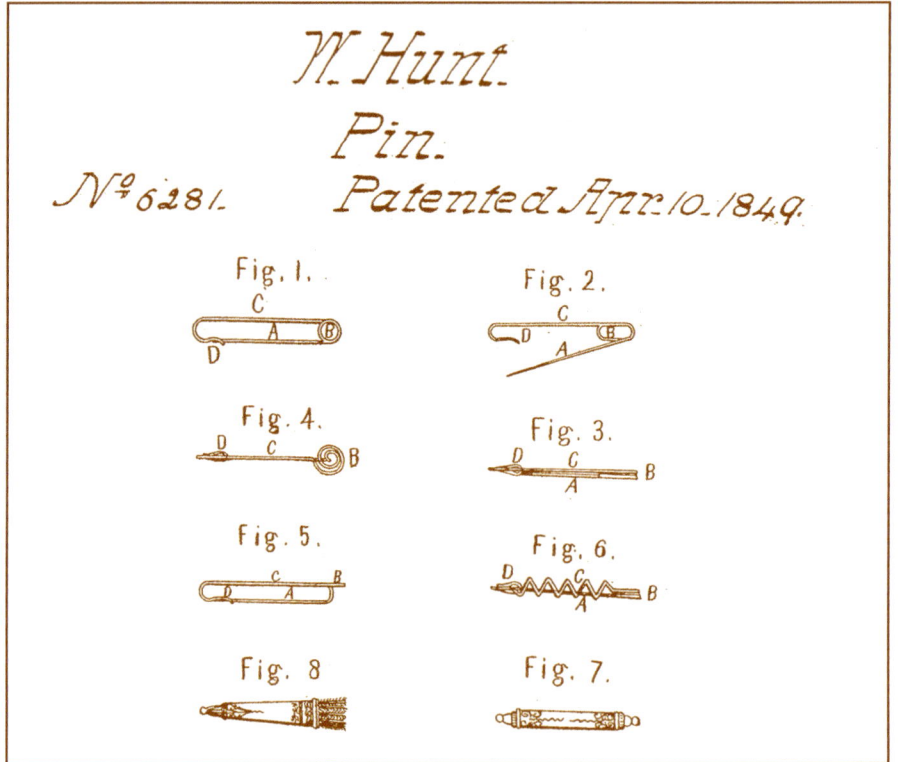

Illustrations from Hunt's 1849 patent for the dress pin. Figures 1 and 2 show two versions of the pin from the side, and figures 3, 4 and 5 show these pins from above. Hunt notes that it would be possible to attach decorations to the bar (labeled C); figures 6 through 8 are examples of decorated pins.

sewing machine could produce only straight seams, and only for a few inches at a time.) In 1846, Elias Howe patented a similar lockstitch sewing machine. A few years later, Isaac Singer infringed on Howe's copyright. In the legal battle that ensued between Singer and Howe, the court decreed that Hunt had invented the eye-pointed needle, but nothing more. Singer was left to reap the rewards and riches of the sewing machine, which has often been called one of the most important inventions in history.

TWISTED LOGIC

The sewing machine was not the only time Hunt let a potential fortune slip through his fingers. Fifteen years later, he invented and patented the safety pin, but then sold his rights to it for next to nothing.

Of the stories that still circulate about Hunt, the most famous is that of the safety pin. Hunt needed to pay off a $15 debt. While he pondered his options, he twisted an eight-inch (20-cm) length of brass wire. After three hours, he stared at the bent piece of metal and realized he had created something useful—a new type of pin.

Hunt had coiled the wire so that it had a spring in the middle. He created a "safety" clasp at one end of the wire. This design, which by now is familiar to almost everyone, allowed the remaining sharp end of the wire to be forced by the spring into the clasp. The tension between the spring and the clasp made the pin strong, and the clasp itself kept whoever was wearing the pin from being pricked. In the patent application Hunt wrote of "the perfect convenience of inserting these into the dress, without danger of bending the pin, or wounding the fingers."

Hunt's safety pin was obviously not the first device to be invented to hold clothes together. Pins had been used since ancient times—the Greeks and Romans used a type of pin, known as a fibula, to fasten their robes. Hunt's pin was also not the first safety pin. In 1842, Thomas Woodward, another New Yorker, patented a straight pin with a safety shield, which he called a "Victorian shielded shawl and diaper pin." Hunt's improvement lay in the coiled spring.

TIME LINE

1796	1829	ca. 1834	1846	1849	1859
Walter Hunt born near Martinsburg, New York.	Hunt patents a knife sharpener.	Hunt invents the lockstitch sewing machine.	A similar sewing machine is patented by Elias Howe.	Hunt patents the safety pin.	Hunt dies.

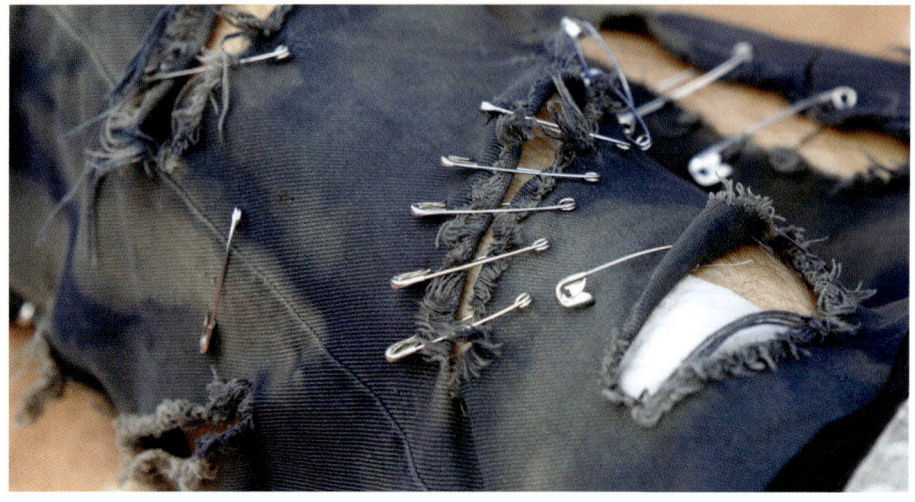

The safety pin has been put to both practical and ornamental uses.

Hunt received U.S. Patent No. 6,281 for what he called a "dress pin" on April 10, 1849. He quickly sold his patent for $400, according to most sources, to pay a debt. One can only wonder if he ever suspected that the safety pin, as it came to be called, might have earned him a fortune. (In October 1849, six months after Hunt, an Englishman, Charles Rowley, patented his own safety pin in England, unaware of Hunt's invention.)

A WEALTH OF INVENTION

Hunt invented many other devices, obtaining 26 patents by the time of his death in 1859. Inventions attributed to Hunt include a streetcar bell, a fountain pen, a tree saw, an ice plow for ships, a shoe heel, road sweeping machinery, machines for making nails and rivets, a revolver, and a type of bullet. A knife sharpener Hunt patented in 1829 is still used.

In his lifetime, Hunt earned mentions in various magazines, including *Atlantic Monthly* and *Scientific American*, but did not create a lasting reputation for himself. Whereas some inventors used marketing savvy to ensure their success, Hunt invented, it seems, for invention's sake.

—Laura Lambert

Further Reading

Kane, Joseph Nathan. *Necessity's Child: The Story of Walter Hunt, America's Forgotten Inventor*. Jefferson, NC: McFarland, 1997.

See also: Cloth and Apparel; Household Inventions; Howe, Elias.

CHRISTIAN HUYGENS

Pioneer in astronomy and clock making
1629–1695

Christian Huygens's name is less well known today than that of his famous contemporary, Isaac Newton (1642–1727). Huygens was, however, a central figure in the development of the scientific fields of astronomy, mechanics, dynamics, and physics. Huygens is credited with many important discoveries. Notable among these was a telescope lens he created that could allow him to see farther and more clearly into space than any previous lens. He also made important contributions to the construction of clocks and proposed a wave theory of light that has contributed significantly to the understanding of light in the 21st century.

EARLY YEARS

Christian Huygens was born in The Hague, Netherlands, on April 14, 1629. His father, Constantin Huygens, a well-known diplomat, scholar, and author, oversaw his son's education until Christian was 16 years old. Among the private tutors he hired for young Christian was the famous French mathematician and philosopher René Descartes (1596–1650), who was a friend of the family. The young student was deeply influenced by Descartes's mechanistic theories of nature, which held that all occurrences in nature are the effect of matter influencing other matter. Descartes was an important influence on Huygens throughout his career.

Although his early education included training in languages, literature, music, and mathematics, Huygens showed particular brilliance in

Undated portrait of Christian Huygens.

mathematics and in creating various types of mechanical models. After his home tutoring, Huygens studied mathematics and law at the University of Leiden in the Netherlands. In 1647, he enrolled at the newly established Collegium Arausiacum (College of Orange) in Breda; he completed his formal education two years later. In 1649 he returned to his father's home, where he continued studying mathematics. He published his first works, mostly concerning mathematical problems, during this period.

TELESCOPES AND ASTRONOMICAL OBSERVATIONS

In 1654 Huygens became interested in telescope design. With his brother, he developed a way of grinding and polishing lenses that allowed him to see much farther into space than any previous telescope had allowed. In 1655 he incorporated these new lenses into a 23-foot (7-m)-long telescope and began to study the sky. At this time he started to reside for lengthy periods in Paris and London, where he demonstrated his work to prominent scientists and attended meetings of the leading scientific societies in those cities.

By this time astronomers had already detected the existence of the planet Saturn. No one understood what accounted for its strange shape, however. Galileo (1564–1642), just one generation earlier, had described

A view from the Cassini-Huygens spacecraft of Saturn's largest moon, Titan, from under the planet's rings of ice. Images taken on April 28, 2006, with red, green, and blue spectral filters were combined to create a natural color view.

Undated portrait (artist unkonwn) of Christian Huygens at work on the first pendulum clock.

Saturn as "three spheres which almost touch each other, which never change their relative positions, and are arranged in a row along the zodiac so that the middle sphere is three times as large as the others." With his improved telescope, Huygens saw that Saturn was not composed of three separate bodies. He detected a thin, flat ring surrounding Saturn that did not touch the planet's surface but, in fact, orbited it. At first astronomers were skeptical. Huygens, however, had charted the changing angle of Saturn's ring over a long period of time and understood how the ring would appear in relation to Saturn and earth as these planets revolved around the sun. After accurately predicting a specific alignment of Saturn's rings for the summer of 1671, Huygens finally proved himself in the eyes of many astronomers of the day.

Huygens also was the first to report the existence of Saturn's largest moon, which he named Titan. He published these findings in 1659 in *Systema Saturnium*. Huygens also noted the existence of a huge, multicolored cloud of gas and dust in the area of the constellation Orion (now known as the Orion Nebula) and made detailed observations of the surface of Mars.

Huygens continued working on improved lenses and eyepieces throughout his life; one type of eyepiece still in use in the 21st century bears his name. In a work entitled *Cosmotheoros: The Celestial Worlds Discovered*, published after he died in 1698, Huygens speculated about the existence of life on other planets. His opinion was that the universe contained many other life-forms, both animal and plant.

INVENTS PENDULUM CLOCK

Throughout his astronomical observations, Huygens realized the necessity of an accurate clock. To time the movements of celestial bodies and thus determine their paths, an observer needed accurate timepieces.

A grandfather clock is one permutation of the pendulum clock invented by Huygens.

Ships at sea also used clocks to chart their courses. With the inconsistent clocks available, sailors often strayed from their intended routes, and many men and ships were lost at sea. While carrying out his work with telescopes, Huygens began working on clocks.

Existing timekeeping devices had operated using a balance mechanism, including a weight that was allowed to fall at a slow speed, turning gears that would move the hands of the clock. Because the rate at which this weight fell was impossible to regulate, clocks were not reliable.

Huygens began to work on an idea that Galileo had proposed some years earlier. Galileo said that the time needed for a swinging pendulum to move from point A to point B was equal to the time it took the pendulum to swing back from point B to point A. He called this effect isochronicity, or equal time. Toward the end of his life, Galileo also theorized that a pendulum could be used to keep accurate time. However, blindness in his old age prevented Galileo from putting this theory into practice.

Huygens tested Galileo's ideas and found that a swinging pendulum could indeed regulate the hands of a clock far more accurately than the previous weight-governed system. In 1656 he devised a pendulum that moved through its arc once every second, and he installed this pendulum in an existing weight-driven clock. The falling weight drove the gears and prevented the pendulum from being slowed by air resistance. For its part, the pendulum regulated the clock hands to within a few seconds' accuracy per day.

In 1658 Huygens published an article, "Horologium"; he published a much-expanded version in 1673, *Horologium Oscillatorium*, perhaps his most important work. In it he describes, in addition to the operation of his clocks, the principles of motion and momentum that underpinned his invention. Within just a few years, Huygens's pendulum clock, which includes the familiar grandfather clock, had become popular throughout Europe.

WAVE THEORY OF LIGHT

Later in life, Huygens turned his attention to questions that many scientists of his day were debating: What is the nature of light? By what mechanism did objects in the world make themselves visible to the human eye? This problem had occupied the minds of scientists and philosophers for millennia. The ancient Greeks believed that some force within the eye was sent out to "feel" objects. Another theory held that objects in the world sent out something that hit the eye. This theory was called the "emission" theory because it supposed that objects emitted some kind of force outward in all directions, making themselves capable of being perceived.

Huygens proposed another theory of light. He postulated that light was composed of a series of waves, similar to sound waves, that traveled through the space between an object and the human eye. This space, he suggested, was filled with an invisible substance, which Huygens called "ether." The waves of light would pass across this ether, and as each wave expanded it became a new source continuing to radiate the same frequency of light. Each particle of the ether, therefore, was seen as the source of another wave of light. The light wave thus altered the

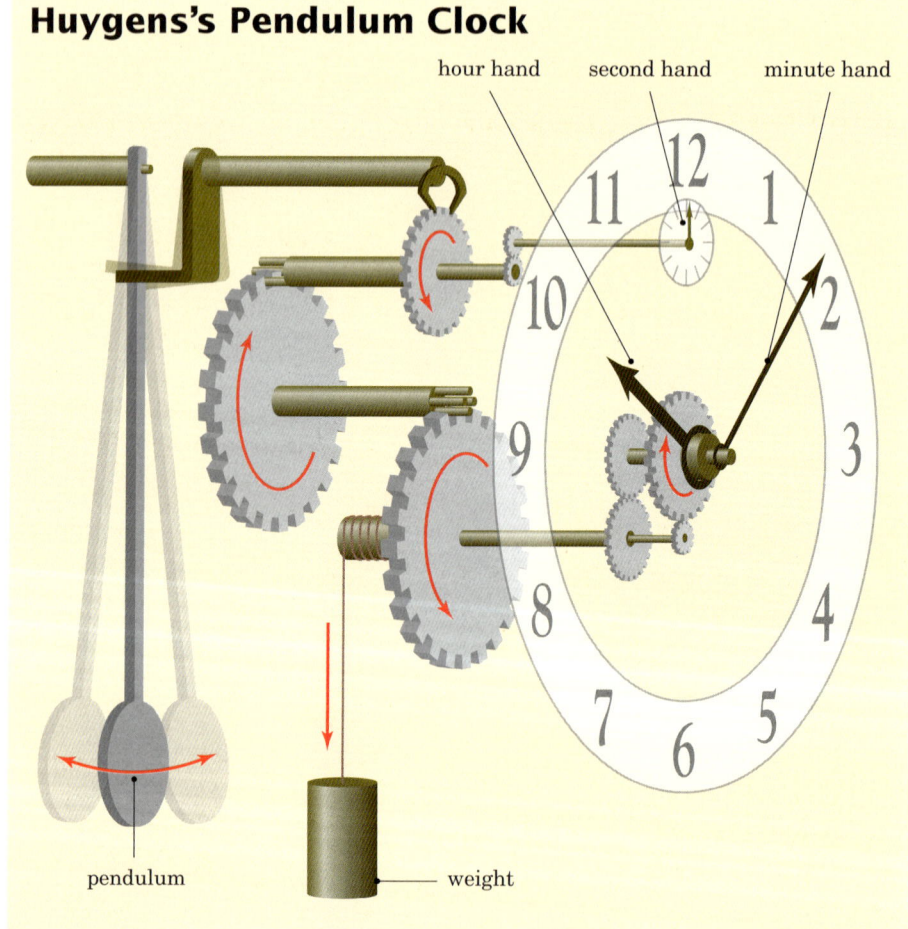

Huygens realized that the time it took for a pendulum to swing from one side of the clock to the other was dependent on the length of the pendulum itself. Once constructed, the pendulum clock was the most accurate clock up to that point in history.

medium between the object and the perceiving eye until it finally affected the eye itself. Huygens's theory also became known as the undulatory theory because it imagined light waves undulating between objects and viewers. Huygens published his account of this theory, *Treatise on Light,* in 1690.

Huygens's ideas about the behavior of light were ill-received at first. Although some scientists found merit in the concept of light waves undu-

Galileo, Huygens, and the Pendulum

Christian Huygens is credited with inventing the pendulum clock, which vastly improved the accuracy with which clocks at sea or on land could keep time. Huygens's first clocks were accurate to within one minute a day. Later, his improvements reduced that error to about 10 seconds a day.

Huygens got the idea for a pendulum clock from his predecessor Galileo Galilei. The great Italian scientist and inventor first started thinking about the motion of pendulums while watching a lamp suspended from a cathedral ceiling in Pisa, where he was a student. Timing the swings by using his pulse, he realized that the complete arc of the lamp's swing was isochronous—that is, it took the lamp the same amount of time to move from one extreme position to the other, regardless of direction. He began thinking that pendulums could be used to keep time. Galileo later formulated more detailed laws about how pendulums work. However, he never used his ideas to design a clock.

When Huygens began working with Galileo's ideas, he found that a pendulum could maintain isochronicity only over a small amplitude (the distance of the pendulum's arc). He also discovered that a truer isochronicity could be achieved if the pendulum swung in a motion that was not quite circular.

Huygens incorporated his improved understanding of pendulums into his clock designs. His clocks caught on quickly as people began to appreciate their superior accuracy. Soon grandfather-style clocks were appearing throughout the Netherlands and then all of Europe.

lating from objects to the perceiving eye, most were very skeptical about the idea of an "ether" that was altered by such waves. Huygens's ideas fell into disregard until the 19th century, when Thomas Young, a scientist, reformulated Huygens's wave theory, omitting the notion of ether.

Along with his telescope and pendulum clock, Huygens is credited with developing the law of conservation of momentum. According to this law, which he derived from his work with pendulums, the energy of a moving object remains unchanged until the object either is stopped or changes direction. He made many other important contributions to the study of mechanics and is considered to be one founder of the science of dynamics, which examines the behavior and impact of moving bodies.

LAST YEARS

After years of living and working in Paris and London, Huygens returned to his native Holland, specifically The Hague, in 1681. Declining health forced him to remain there until his death on July 8, 1695. He lived mostly alone, never marrying. He also chose not to take students—another reason his ideas did not receive greater circulation during his lifetime and immediately after his death.

Christian Huygens was little known at the time of his death. Despite his lifetime of important scientific research, his accomplishments were overshadowed by those of Newton. Later, thanks to Young and other scientists, Huygens's ideas began to gain greater prominence. By the 20th century he was rightfully considered to be one of the greatest scientists and inventors of the seventeenth century, a period that saw an unparalleled explosion of scientific inquiry and experimentation.

The Cassini-Huygens Program, a joint effort of the National Aeronautics and Space Administration (NASA), the European Space Agency, and the Italian Space Agency, was launched in October 1997 to study Saturn and its moons. Both the overall program and a specific atmospheric probe of Titan were named after the great Dutch scientist who first identified Saturn's largest moon and its mysterious rings. The unmanned

TIME LINE

1629	1647	1649	1655	1656
Christian Huygens born in The Hague, Netherlands.	Huygens enrolls at the Collegium Arausiacum (College of Orange).	Huygens returns home and continues studying mathematics.	Huygens designs telescope and begins studying astronomy.	Huygens designs a pendulum clock.

The launch of a Titan IVB/Centaur spacecraft, carrying the Cassini orbiter and its Huygens probe, on October 15, 1997. The craft reached Saturn in July 2004 after a 2.2-billion-mile (3.5-billion-km) journey.

TIME LINE

1659	1671	1673	1698	1695
Huygens publishes *Systema Saturnium*.	Huygens accurately predicts a particular alignment of Saturn's rings.	Huygens publishes *Horologium Oscillatorium*.	*Cosmotheoros: The Celestial Worlds Discovered* is published.	Huygens dies.

Cassini-Huygens spacecraft arrived at Saturn in July 2004 to begin a four-year exploration, providing dramatic flyby images of both Saturn and Titan.

—Paul Schellinger

Further Reading

Book
Bell, A. E. *Christian Huygens and the Development of Science in the Seventeenth Century.* London: Edward Arnold, 1947.

Web site
Christiaan Huygens
  Extensive collection of Huygens resources from the University of St. Andrews, Scotland.
  http://www-history.mcs.st-andrews.ac.uk/Biographies/Huygens.html

See also: Hooke, Robert; Newton, Isaac; Science, Technology, and Mathematics.

INVENTION AND INNOVATION

People tend to think of invention in terms of a single moment when an inventor suddenly gets the inspiration to solve a long-standing problem. Inventions, however, can come into the world in many ways. Blue jeans were born as a new idea and put on the market within two years of their invention, whereas computers have developed through gradual improvements spanning more than three centuries. Even fully formed inventions are not guaranteed immediate success: a great idea can change the world overnight, or it may take years to win over a skeptical public. Innovation is the process by which an invention arrives in the world, captures people's imaginations, and becomes successful, sometimes through the involvement of many different inventors.

A general view of the Tokyo Game Show 2006 in Makuhari, Japan. Companies attend trade fairs like this one to show off their new and improved products.

WHAT DRIVES INNOVATION?

An idea is the starting point for an invention and the process of innovation through which it develops. Inventors get their ideas in many different ways, sometimes after a long and deliberate search for inspiration and sometimes out of the blue. As a boy, Chester Carlson (1906–1968) was determined to escape a life of hardship and poverty by making a great invention. More than twenty years later, after a great deal of research and experimentation, he figured out his winning idea: the office photocopier. The inventor of the safety razor, King C. Gillette (1855–1932), also began his career by looking for an idea that would make money. After a friend challenged him to think of something simple that customers would have to buy repeatedly, he conceived of the disposable razor blade.

In the modern age, most inventions appear because large corporations—3M, DuPont, IBM, and Lucent, for example—employ scientists and engineers specifically to devise new products. A brilliant young chemist, Wallace Carothers (1896–1937), invented nylon, the first synthetic plastic fiber, after the DuPont Chemical Company lured him from his academic job at Harvard. Once the idea of synthetic fibers had been born, chemists developed many other revolutionary materials using the same process. In this case, innovation was driven by companies seeking new products and greater profits.

Innovation can also be driven by customers who feel companies are not making the products they want. Betty Nesmith Graham (1924–1980) was a secretary frustrated by the laborious erasing needed to correct her typing mistakes. When IBM introduced a new range of typewriters with a different kind of ink, erasing mistakes became almost impossible; thus entire pages often had to be retyped. So Graham invented a quick-drying white fluid—Liquid Paper—to cover typing errors. After IBM showed no interest in her invention, she decided to manufacture it herself and became a multimillionaire.

Some inventors make innovations by combining a talent for tinkering with a genius for science. The brothers Orville (1871–1948) and Wilbur (1867–1912) Wright were mechanics who perfected their flying machines through careful scientific experiments. John Bardeen (1908–1991) was an outstanding theoretical scientist who helped to invent the transistor after meeting Walter Brattain (1902–1987), a brilliant experimenter. For others, tinkering becomes a kind of science, in which an invention gradually evolves through trial and error. Les Paul (1915–) developed the solid-body electric guitar by experimenting with many different gadgets and materials until he achieved just the sounds he wanted.

> If I have seen further, it is by standing on the shoulders of Giants.
> —Isaac Newton

DEVELOPING OTHER PEOPLE'S IDEAS

Not all inventors come up with ideas of their own. Innovations often happen

Marketing Myopia

An American academic, Theodore Levitt, inspired many companies to innovate after he published a classic essay, "Marketing Myopia," in the *Harvard Business Review* in 1960. The basic premise was that companies often go out of business because they are shortsighted (myopic) and do not see how times are changing. Levitt argued that the great steam railroad companies had gone out of business precisely because they saw themselves as "steam railroad companies." If they had defined themselves more broadly, as being in the transportation business, they might have become pioneers of truck or airplane freight and changed with the times before these newer technologies made much of their business obsolete.

Overcoming such myopia can lead to groundbreaking innovations. In the late 1960s, the photocopier company Xerox realized computers could make offices "paperless," which would threaten its business. This prompted Xerox to change the way it saw itself: not narrowly, as a photocopier manufacturer, but more broadly, as a company that made machines for processing information. Following this decision, Xerox set up a new research laboratory at Palo Alto (Xerox Palo Alto Research Center, or PARC) where many of the innovations behind modern personal computers were developed in the early 1970s. In much the same way, oil companies have increasingly defined themselves as "energy suppliers" and begun to move into technologies such as solar energy and wind power in preparation for the time when oil runs out.

---

when inventors build on ideas or discoveries others have made. Many people believe U.S. tailor Levi Strauss (1829–1902) invented blue jeans. In fact, another tailor, Jacob Davis, had the idea to reinforce denim work pants with copper rivets and sought Strauss's help afterward. Strauss contributed business and marketing acumen; these talents, as much as the idea itself, are what made blue jeans such a success. The inventors of Rollerblades, Scott (1959–) and Brennan Olson (1963–), also built on an existing idea. After discovering a long-forgotten type of roller skate in a sports store, they bought the patent, made many refinements, and advertised their product themselves by skating around town. Their innovation was to reinvent and publicize a great idea that had been forgotten.

Some inventions are developed by a whole series of inventors over many years without much success, until someone

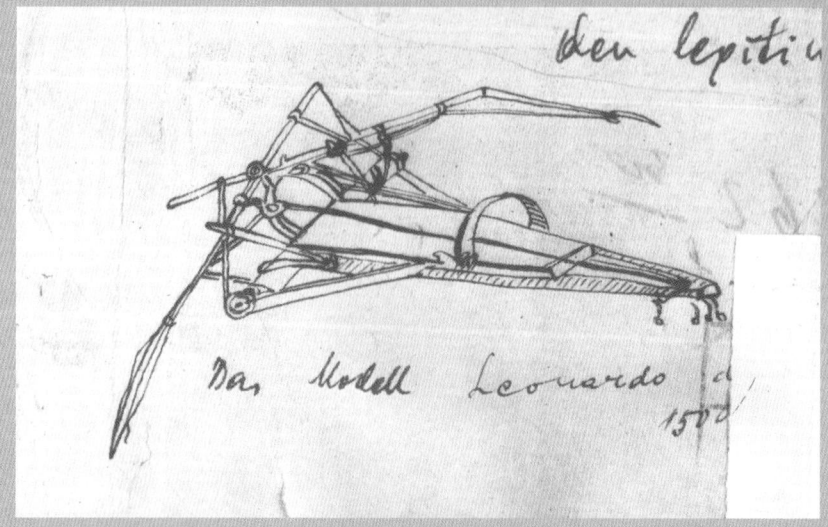

A sketch of a flying machine by Leonardo da Vinci from around 1500; the idea of human flight was an old one, but it took the Wright brothers to turn the idea into reality.

makes a crucial innovation. Many people had tried to invent a sewing machine, but Elias Howe (1819–1867) made the all-important breakthrough: his machine had a needle with a hole at the bottom and two threads that could "lock" the material together. Flying machines are another example: historians think people have been trying to fly since ancient times, but never found a reliable way of getting off the ground and staying aloft. When the gasoline engine was invented in the late 19th century, the Wright brothers saw how they could use it to power a glider, making the crucial innovation that led to the airplane.

Often so many individuals are involved in an invention that it cannot be considered to have a single inventor. Some say George Fuller (1851–1900) invented the skyscraper, but architect William Le Baron Jenney (1832–1907) had already made a tall building with a steel framework and many others also played a part; elevators, developed by Elisha G. Otis (1811–1861), and escalators, invented by Jesse Reno (1861–1947), were vital also. The computer is another example of an invention with no single inventor. For hundreds of years, innovative scientists, mathematicians, and engineers built on one another's work to produce the computers people use today—but the computer remains a "work in progress" for which no individual can claim sole credit.

PUTTING IDEAS INTO PRACTICE

As for innovation, thinking up an idea—or building on someone else's—is only part of the story. Putting the idea into practice by successfully manufacturing a product, selling it at the right price, and advertising it so people know about it are also crucial. Even the greatest invention will fail commercially if it is too expensive or if, because of poor advertising, no one knows it exists.

What characterizes the best innovators is their ability to turn great ideas into products people want. English scientist Michael Faraday (1791–1867) made the first electricity generator in 1831, but his machine was just a small piece of funny-looking laboratory equipment. The marketing genius of Thomas Edison (1847–1931) was needed to build huge generators and power plants, and to popularize home lighting so people would want to buy his electricity. Karl Benz (1844–1929) and Gottlieb Daimler (1834–1900) invented the automobile. Yet American industrialist Henry Ford (1863–1947) made the crucial innovation when he found a way to mass-produce cars, making huge quantities that were inexpensive enough for most people to buy. English scientist Oliver Lodge (1851–1940) was one of the inventors of radio and accused his Italian rival Guglielmo Marconi (1874–1937) of "inventing nothing and borrowing everything else." Like Edison, Marconi was an innovator who put the science of radio into practice. His bold demonstrations (for example, showing how radio could carry messages across entire oceans) captured people's imagination in a way Lodge's laboratory experiments never could.

EVOLUTION AND REVOLUTION

Some innovations help to reinforce other people's ideas by making them better. This kind of innovation is sometimes considered incremental (adding to what is already there), sustaining (allowing devices to remain essentially the same but with a slight improvement), or evolutionary (helping existing technology to develop). Other innovations change the world more radically. When German craftsman Johannes Gutenberg (ca. 1400–1468) invented the printing press, books could be produced in numbers never before possible and hand-copying of books became obsolete. This type of innovation is considered revolutionary (a total transformation) or disruptive (the invention totally redirects how a process is carried out).

Most modern inventions are developed by large corporations that innovate slowly and incrementally over a long period. Every few years, Microsoft releases a new version of its Windows software. Each version has improvements over the last one but is "upwardly compatible," meaning that existing users can move across without too much disruption. Most automobile companies work this

In 1901 a colleague of Guglielmo Marconi prepares a kite to support an aerial at Signal Hill station, in Canada, to receive the first transatlantic wireless signal from Cornwall, England. Bold demonstrations like these helped excite public interest in telegraphy.

Microsoft CEO Steve Ballmer, far right, talks to reporters about the latest release of the Windows operating system, called Vista, in 2007.

way, too, bringing out slightly improved models each year to encourage customers to keep buying. In the future, an inventor might develop a radically new form of computer that makes Microsoft's products obsolete. Someone could invent a flying car, immediately consigning today's cars to oblivion. Invention and innovation advance slowly and steadily, mostly by a gradual process of evolution, but the gigantic leaps forward—the revolutionary breakthroughs—are what we tend to notice and remember.

—Chris Woodford

Further Reading

Books

Von Oeck, Roger. *A Whack on the Side of the Head*. New York: Warner Business, 1998.

Wulffson, Don. *The Kid Who Invented the Popsicle, and Other Surprising Stories about Inventions*. New York: Puffin, 1999.

Web sites

Innovative Lives

Stories of invention and innovation from the Smithsonian Institution.

http://www.invention.smithsonian.org/centerpieces/ilives/

Invention at Play

Discover how play helps creativity and invention.

http://www.inventionatplay.org/

See also: Bardeen, John, Walter Brattain, and William Shockley; Benz, Karl; Carlson, Chester; Carothers, Wallace; Corporate Invention; Edison, Thomas; Faraday, Michael; Ford, Henry; Fuller, George; Gillette, King C.; Gutenberg, Johannes; History of Invention; Howe, Elias; Marconi, Guglielmo; Olson, Brennan, and Scott Olson; Otis, Elisha; Paul, Les; Reno, Jesse; Strauss, Levi; Wright, Orville, and Wilbur Wright.

JOSEPH-MARIE JACQUARD

Inventor of the Jacquard loom

1752–1834

Joseph-Marie Jacquard grew up in an era and a community where silk weaving was tremendously important to the economy. He worked to improve the weaving of silk tapestries, developing an invention that is still used today to speed the production of elaborately patterned fabrics. Jacquard was widely celebrated during his lifetime for his contributions to the textile industry. After his death, however, his invention had an application that he had most likely never anticipated: his punch-card system for programming a loom was used to program computers until well into the 20th century.

EARLY YEARS

Joseph-Marie Jacquard was born in Lyon, France, in 1752. At the time, Lyon was the center of silk weaving in France. Jacquard's father was a weaver of silk brocades—fabrics with raised designs. The family appears to have slid into and out of poverty. Joseph-Marie and his older sister were the only two of the nine Jacquard children to survive into adulthood; his mother died in 1762, when he was 10.

The young Jacquard did not go to school; he helped out in his father's workshop instead. He did not learn to read until he was 12, with the help of his sister's husband. In 1772, Jacquard's father died, and Jacquard inherited a modest fortune. He married in 1778 and his son, his only child, was born a year later. Jacquard worked as a weaver, but his business did not prosper. By 1783 he had spent almost his entire inheritance, and his wife had to work in a factory to help support the family, while Jacquard took a variety of menial jobs.

Undated portrait of Joseph-Marie Jacquard.

THE FRENCH REVOLUTION

Jacquard's life changed dramatically during the French Revolution, which began in 1789. Lyon's economy was dependent on luxury goods bought by the aristocracy; accordingly, many of Lyon's people opposed the revolution. The city was besieged by revolutionary forces in 1793. Jacquard and his son, then 15 years old, fought on the royalist side; Jacquard in particular earned a reputation for great bravery during the siege.

Lyon fell, however, and Jacquard and his son fled. They joined the revolutionary forces under assumed names. Jacquard was quickly promoted to officer, but his son was killed in combat, most likely in 1797. Jacquard himself was badly wounded and returned to Lyon to be hospitalized in 1798. The following year, Napoleon Bonaparte took over France, putting an end to the chaos of the revolution.

IMPROVING LOOMS

At some point following his return to Lyon, Jacquard recovered from his wounds and began trying to improve the silk loom. Woven silk, especially silk with elaborate brocades or designs, was in demand both in France and abroad. However, weaving designs into silk was an extremely time-consuming process, and silk weavers often could not begin to meet the demand for their products.

> You are aware that the system of cards which Jacquard invented [is] the means by which we can communicate to a very ordinary loom orders to weave any pattern that may be desired. Availing myself of the same beautiful invention I have by similar means communicated to my Calculating Engine orders to calculate any formula however complicated.
>
> —Charles Babbage

To weave fabric, a weaver lays out a set of threads, called the warp, onto a loom. The weaver then interlaces a second set of threads, called the weft, at a right angle to the first. In the 1700s, looms had evolved to the point where creating a plain fabric was a task that could be done quickly. If a weaver wanted to create a design, however, the process became more complicated and much slower.

To create a design on a tapestry, specific warp threads had to be lifted up out of each row of weaving. For example, to create a red rose, red warp threads had to be lifted out of the weave at just the right place to lie on top of the weft threads, displaying their red color. Complicated patterns required lifting specific threads to achieve the design changes for each row. Silk thread is quite fine, and it was not uncommon to have about five hundred warp threads in a single row of weaving. Weavers employed workers (drawboys) to lift the threads, but even with their help, the process was extremely slow. A skilled weaver and drawboy using the best equipment available could produce only about an inch (2.54 cm) of designed silk cloth a day.

In late 1800 Jacquard patented an improved loom that could create silk fabric decorated with small, simple patterns without the help of a drawboy. Instead, the weaver used foot pedals to create a limited number of patterns. Jacquard entered his loom in an exhibition on French industry in Paris in 1801. The loom caused a minor sensation; Jacquard was reportedly attacked by drawboys who were worried that his invention would replace them. Jacquard was given an award by the French government for his accomplishment.

A year later, Jacquard entered another loom he had invented, this one for making fishnets, in an industrial contest. He received both an award and a grant of 1,000 francs, enough money to support his work in developing yet another loom.

THE JACQUARD LOOM

In 1804 Jacquard unveiled the loom that now bears his name. The Jacquard loom incorporated ideas from other, experimental looms, but in a way that created a uniquely practical machine.

Like Jacquard's earlier loom, the Jacquard loom could be operated without the help of a drawboy, because a mechanical device that could lift the warp threads had been added. Perhaps the most innovative feature of the Jacquard loom, however, was that it could be "programmed" to create any desired design by using punch cards. A punch card is a small card made of thick paper with a pattern of holes.

An earlier experimental loom had used punch cards to tell a thread-lifting device what warp threads to lift to weave the row correctly. However, that loom had required a drawboy to insert a new punch card for every new row. Jacquard's new loom, in contrast, used a series of punch cards fastened together into a belt. A feeder for the loom would automatically draw the belt forward whenever a row was completed, positioning a new punch card in place for the next row.

The punch-card system endowed the loom with flexibility; the punch cards could be arranged to create a repeating pattern or to weave one large design. Unlike Jacquard's earlier loom, this loom had no limitations on the type of design that could be automatically woven. In addition, if a weaver wanted to use an identical design in another tapestry, the cards for that design could be reused. Since the Jacquard loom lifted the correct threads automatically, the process of weaving designs became much faster. A weaver using a Jacquard loom could weave up to two feet (.6 m) of fabric each day, in any design imaginable, and all without the help of a drawboy.

Jacquard's loom astonished his contemporaries, and Jacquard himself received enormous praise. In 1805, Napoleon decreed Jacquard's loom public property, guaranteeing its inventor a lavish annual pension

How the Jacquard Loom Works

To use a Jacquard loom, a weaver first created a design and then determined how the threads would lie in each row. The thread design for each row was then translated into a pattern of holes in a punch card.

The punch cards were attached together to form a belt, which was then fed into the loom. Each time a row was woven, the loom would pull a new card into place. The card was then pushed up firmly against the ends of a block of spring-loaded rods. Each rod was attached to a hook, which was attached to a particular warp thread. If a rod came up against solid card stock, it would remain in place, as would the hook it controlled, and the warp thread attached to the hook would be woven into the row normally.

If a rod came up against one of the holes punched into the card, however, the rod would slide forward. The sliding action would also move the hook forward so that it hooked onto a bar. The bar would rise, lifting the hook and the attached warp thread out of the weave for that particular row. Once the row was completed, the bar would drop, releasing the hooks. The loom would then draw in the next card and the entire process would repeat.

A Jacquard loom from 1825.

Chromolithograph of a Jacquard power loom from 1915. The punch cards are visible hanging off the top of the loom.

from the French government and a generous royalty for every Jacquard loom brought into use.

By 1812, around eleven thousand Jacquard looms were in use in France. The looms allowed French weavers to make much larger, more elaborate designs than had been practical with previous looms. The result was a boom in the French textile industry and a sharp increase in demand for French fabrics abroad. France tried to maintain this advantage by keeping the new technology secret, but details of the Jacquard loom eventually leaked out. By the 1830s, Jacquard looms were in use in other European countries, including Great Britain.

Jacquard continued making improvements to his loom. In 1819, he received the Cross of the Legion of Honor, one of France's most prestigious awards. In the 1820s, Jacquard retired to the village of Oullins, where he died in 1834.

Jacquard looms are still in use, and the word jacquard is now a common noun meaning a fabric with an elaborate weave or pattern. Modern Jacquard looms are powered by electricity, and computers are used to develop the designs, but the basic process remains unchanged.

TIME LINE

1752	1772	1783	1789	1800
Joseph-Marie Jacquard born in Lyon, France.	Jacquard's father dies, leaving a modest fortune.	The fortune spent, Jacquard performs menial labor while his wife works in a factory.	The French Revolution begins.	Jacquard patents an improved loom.

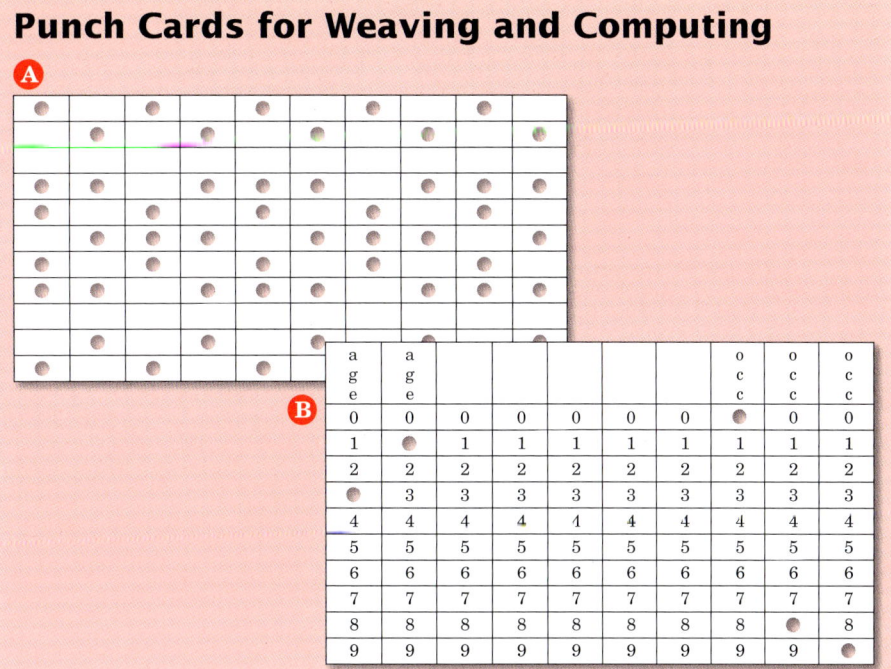

Punch Cards for Weaving and Computing

Jacquard's punch card could be used for very different purposes, as these schematic drawings show. The top card (A) is an example of instructions to a loom for a pattern to be woven in a piece of cloth; the bottom (B) is a sample of data sent to U.S. Census Bureau tabulating machines.

PROGRAMMABLE MACHINES

Jacquard's legacy, however, does not end with the textile industry. In the 1830s, a British scientist, Charles Babbage (1792–1871), attempted to build something he called the Analytical Engine, a precursor to the modern computer. In the process, he envisioned many concepts that would later be fundamental to computing. One of these ideas was that of programming. Babbage realized that the punch-card system Jacquard had developed to communicate designs to his loom had a broader use: to communicate instructions to other kinds of machines, including the Analytical Engine.

TIME LINE

1802	1804	1811	1819	1834
Jacquard receives an award and a grant to support his work.	Jacquard introduces the Jacquard loom.	About eleven thousand Jacquard looms are in use in France.	Jacquard is awarded the Cross of the Legion of Honor.	Jacquard dies.

A runway model wears a coat made of jacquard fabric, named in honor of the inventor.

Although the Analytical Engine would never become practical, punch cards were used in a variety of other machines that had to process large quantities of information, including the first tabulating machines used by the U.S. Census Bureau for the 1890 census. In the 1940s, the first modern computers developed in the United States were programmed with punch cards. Punch cards became the dominant means of programming computers and storing information until the 1970s, when they were replaced by magnetic tapes and disks.

—Mary Sisson

Further Reading

Book
Essinger, James. *Jacquard's Web: How a Hand Loom Led to the Birth of the Information Age.* Oxford: Oxford University Press, 2004.

Web sites
Babbage
    An exhibition on Charles Babbage from the Science Museum in London.
    http://www.sciencemuseum.org.uk/on-line/babbage/index.asp
The Pattern Loom by Joseph-Marie Jacquard
    A presentation on the Jacquard loom by the Deutsches Museum.
    http://www.deutsches-museum.de/ausstell/meister/e_web.htm

See also: Babbage, Charles; Cloth and Apparel; Computers.

EDWARD JENNER

Inventor of the smallpox vaccine

1749–1823

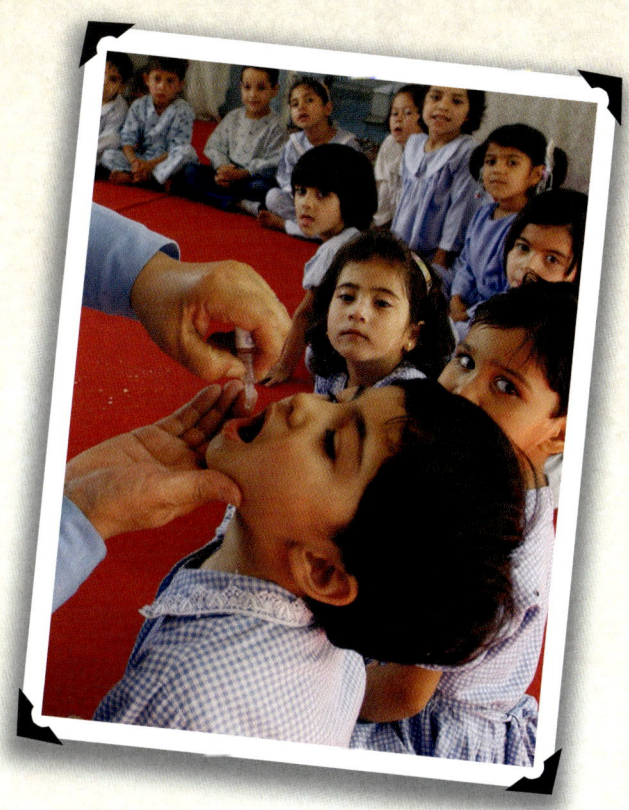

Edward Jenner developed a vaccine for smallpox, which was one of the deadliest diseases affecting England and other countries across the world during the 18th century. Jenner also coined the term vaccination and laid the foundation for modern principles of immunology.

EARLY YEARS

Edward Jenner was born May 17, 1749, in Berkeley, Gloucestershire, in the southwest of England. Growing up in this rural environment, Jenner became a very close observer of the natural world around him. His formal schooling began at Wooten-under-Edge in Gloucestershire, then continued at Cirencester. At age 13 he was apprenticed to a doctor named Daniel Ludlow. In 1770, Jenner went to London to study under the renowned surgeon, anatomist, and naturalist John Hunter at St. George's Hospital. Hunter chose Jenner partly on the basis of the young man's keen interest in botany and zoology, in addition to his medical talents. Jenner studied and worked with Hunter for three years before returning to his native Berkeley in 1773 to establish a practice as a country doctor. He remained in Berkeley, working as a doctor and naturalist, for the remainder of his life.

John Hunter had encouraged Jenner's natural curiosity, urging his bright young pupil to take an imaginative and experimental approach to medicine (while still observing the day's standard medical principles). Jenner soon found occasion to apply his imaginative and observational powers to a question that had long occupied naturalists. He gained the notice of the scientific community by providing an explanation for the strange nesting habits of the cuckoo, which lays its eggs in another bird's nest, expelling the host bird's own young (see box, The Strange Habits of the Cuckoo).

Undated portrait of Edward Jenner.

Jenner's paper detailing his observations of the cuckoo won him election into the prestigious Royal Society in 1789. However, his reputation rests on other work of far greater importance. Jenner's experiments stemming from his observation of the relationship between cowpox and smallpox led him to develop the vaccine that would eventually lead to the elimination of smallpox throughout the world.

A BRIEF HISTORY OF SMALLPOX

By the time Jenner began to study smallpox, the disease had already plagued humans for centuries. After spreading throughout Europe in the sixteenth and seventeenth centuries, smallpox was carried to the Americas during the "voyages of discovery." There, it killed more of the indigenous populations than did all the battles with European settlers. In Jenner's time, smallpox accounted for one-third of all children's deaths in Europe, and it devastated populations around the globe. Caused by the virus variola, which enters the body through the lungs and then spreads through the blood to infect the internal organs, smallpox becomes evident when it erupts as small pink spots that grow to become raised, open blisters.

The Strange Habits of the Cuckoo

Although Edward Jenner is best known for his development of a vaccine to immunize people against smallpox, his interests were wide-ranging. He made important contributions to the field of natural history as well as medicine. The first of these was a paper in which he explained the bizarre nesting habits of the cuckoo, a bird that had puzzled observers for centuries.

An adult female cuckoo will lay a single egg in another bird species' nest, usually that of a hedge sparrow. The host bird's own eggs and fledglings are removed from the nest, and the cuckoo fledgling is left alone to receive the care of the foster parent. Scientists always believed that the adult cuckoo was responsible for clearing the nest prior to laying its one egg there.

Jenner hypothesized that in fact the newly hatched cuckoo fledgling, not the parent, pushed the eggs or young sparrows out of the host bird's nest. One of his findings was that the young cuckoo's body had a unique depression in its back, between its wings, with which it could cup objects and push them backwards. This depression disappears by the time the fledgling cuckoo is 12 days old. Jenner conducted many experiments to support his hypothesis before presenting his findings in 1788 to the Royal Society, which made him a fellow the following year for his contribution.

Jenner's hypothesis remained open to question until the 20th century, when naturalists developed the means of photographing the fledglings in action, proving that Jenner was right.

Historical Smallpox Outbreaks

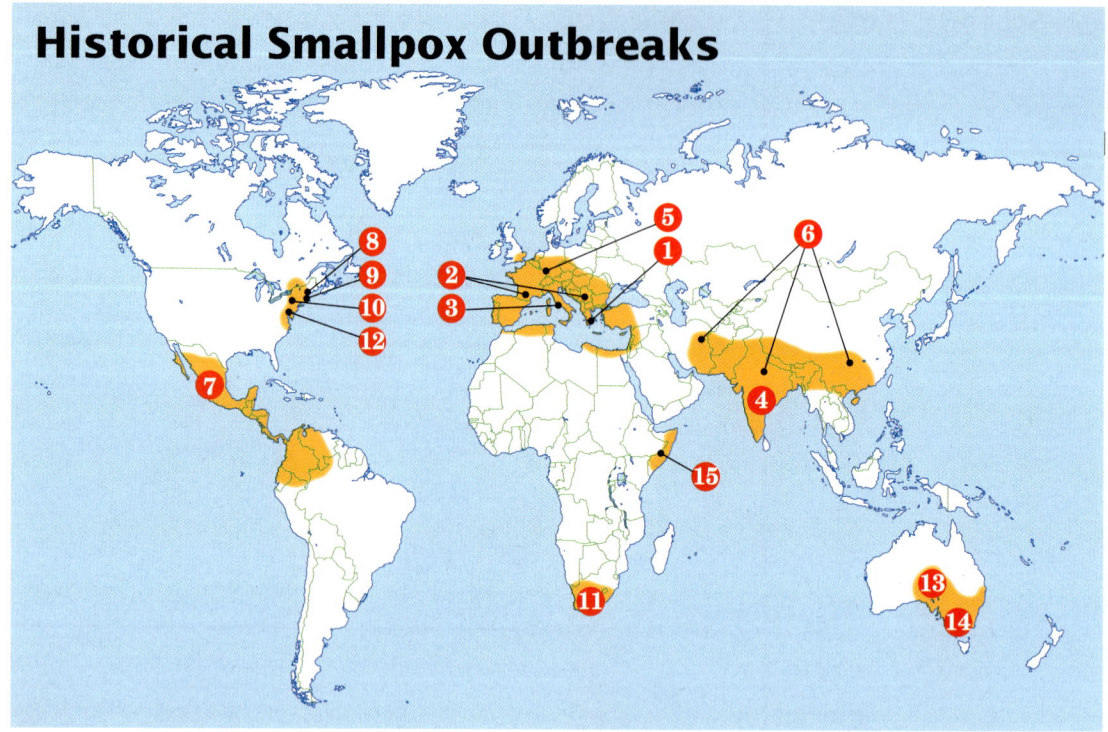

1 Athens—430 BCE
About one-third of the population reportedly died of smallpox during the so-called plague of Athens.

2 Roman Empire—165–180 CE
Historians believe the Antonine plague was either smallpox or measles.

3 Roman Empire—251–266 CE
Historians believe the plague of Cyprian was either smallpox or measles.

4 India—400 CE
Smallpox outbreak.

5 Europe—581 CE
Details about this widespread European epidemic have largely been lost.

6 China, India, and the Middle East—ca. 1000
Faced with deadly smallpox outbreaks, people in China and India began to practice variolation, a practice that spread to the Islamic world.

7 Aztec Empire—1521
About one-quarter of the population of the Aztec Empire was killed after smallpox was brought by the Cortés expedition. Smallpox continued wreaking havoc on native populations; some historians estimate that the total number of deaths due to smallpox was 60 to 90 percent of the total native populations of North and South America.

8 Plymouth Colony—1633
Native Americans in the vicinity of Plymouth colony contracted smallpox.

9 Boston—1636–1721
Boston endured multiple smallpox epidemics.

10 Iroquois Nation—1679
Millions of Iroquois contracted smallpox and died.

11 Cape of Good Hope—1713
The Khoisan people were decimated in a smallpox epidemic.

12 American Colonies—1775–1782
About 125,000 people died of smallpox during the course of the Revolutionary War.

13 Sydney—1789
A smallpox epidemic broke out among the Aboriginal people around Sydney, Australia.

14 Australia—1820s–1830s; 1860s–1870s
Explorers moving inland noticed various smallpox epidemics strike the Aboriginal population.

15 Somalia—1977–1978
Last known smallpox epidemic; last known case in Somalia in 1978.

The Chinese had developed a method of providing immunity against smallpox by blowing flakes from a smallpox scab up the nostrils of healthy people. Later, the practice of variolation—injecting smallpox material under an unaffected person's skin to induce a mild case of the disease that would then leave the person immune—was used by physicians in Greece and Turkey. Lady Mary Wortley Montagu observed this practice in Turkey around 1720 and introduced it to England. Variolation proved effective in some cases, but it was still very dangerous. Many people treated in this manner developed full-blown smallpox.

An undated illustration of Jenner inoculating James Phipps with cowpox.

Around Jenner's time, variolation was used by doctors in Turkey and Greece; they had discovered that injecting some of the pus from small pox blisters into a healthy body could produce a mild case of the disease but leave the body immune afterward. This version of variolation also was very risky. Nevertheless, it became widespread in England by the 18th century, as fears of the disease led people to take extreme measures.

DEVELOPING A VACCINE

Jenner himself had been variolated as a boy and suffered badly as a result. He vowed to eradicate the disease. He became interested in reports that milkmaids who had been exposed to cowpox (similar to smallpox, but much less severe) appeared to be immune to smallpox. After years of observing people who had contracted cowpox and theorizing that there was a connection, Jenner decided to carry out a formal experiment. In May 1796, he extracted fluid from an open cowpox on the hand of Sarah Nelmes, a milkmaid, and used it to inoculate an eight-year-old boy named James Phipps. As Jenner expected, the boy developed cowpox. Then, six weeks later, he inoculated Phipps with smallpox. To his great relief, Jenner

An anti-vaccination cartoon by satirist James Gillray shows people turning into barnyard animals after being vaccinated.

saw that the boy did not develop smallpox. A short time later, he introduced smallpox again on Phipps's skin, and again no infection developed. Jenner tentatively concluded that his theory was correct, reasoning that immunity to smallpox could be achieved much more safely by injecting cowpox material into the bodies of uninfected people.

Jenner named his procedure vaccination, borrowing the Latin name for cow, *vacca*, and deriving from it *vaccinia*, or cowpox. Then in 1798 he published the results of his smallpox research in *An Inquiry into the Causes and Effects of the Variolae Vaccinae*. He continued to carry out research and publish his accounts over the next two years. All the evidence confirmed Jenner's initial theories about immunization.

OPPOSITION TO VACCINATION

Nevertheless, Jenner's technique for preventing smallpox faced various kinds of opposition. One of these was from the medical community itself: many doctors had established successful businesses by becoming "variolators"—those who practiced the risky procedure of injecting people with the smallpox virus. Jenner's much safer and more effective treatment threatened these doctors' incomes. Superstition spawned opposition from other quarters: some people feared being treated with materials taken from "God's lowlier creatures." Even those medical practitioners who supported Jenner's findings faced difficulty incorporating his practices. Cowpox was not nearly as widely occurring as smallpox, and doctors often had to obtain samples of cowpox directly from Jenner. Since the people handling these samples usually were the same people handling smallpox, cowpox samples often became contaminated with smallpox. As a result, many people charged that infection with or vaccination with cowpox was no safer than exposure to smallpox.

> The annihilation of the Small Pox, the most dreadful scourge of the human species, must be the final result of [my] practice.
>
> —Edward Jenner

Even so, Jenner's vaccination methods eventually won widespread favor. The British government recognized his work in 1800; by that year more than one hundred thousand people worldwide had received Jenner's vaccine. Demand increased so quickly that Jenner had to develop a means of transporting the vaccine long distances. This he did by drying vaccination material, so that it could be stored in a glass tube for up to three months.

VACCINE CLERK TO THE WORLD

Jenner spent the remainder of his life as consultant and supplier of his vaccine to doctors around the world. He once referred to himself as "the Vaccine Clerk to the World." Although Jenner did not patent his vaccine, preferring to make it as widely available as possible, the British govern-

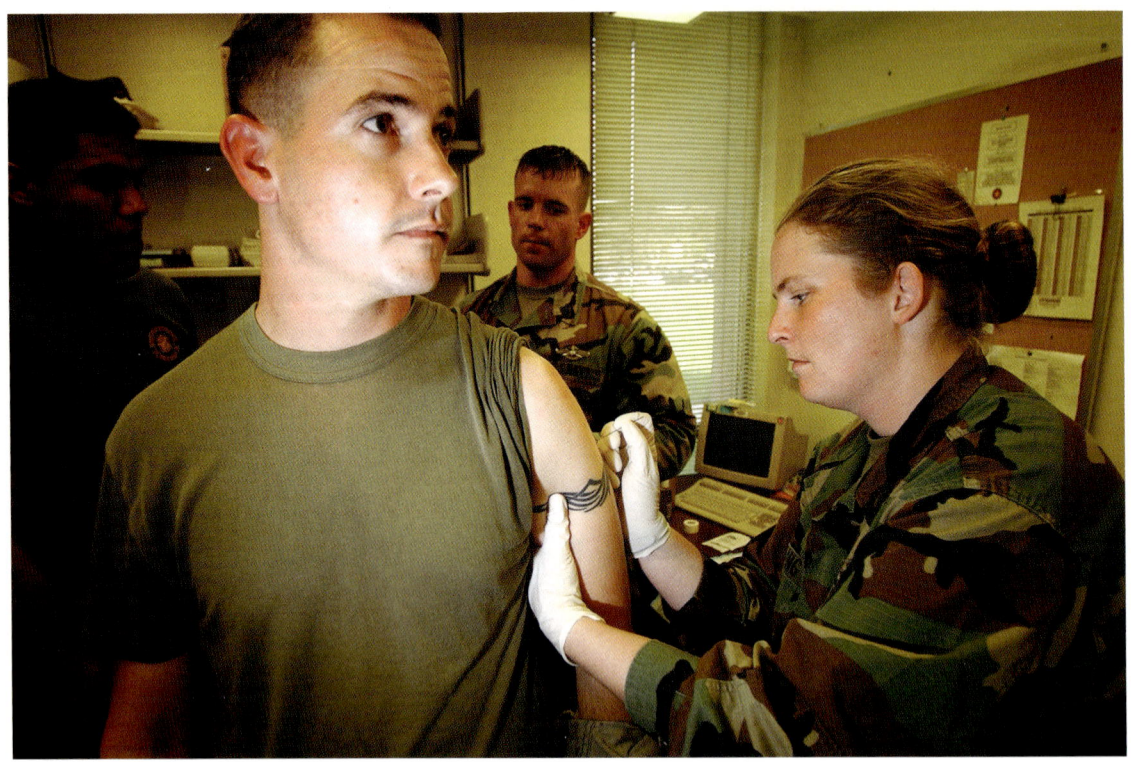

Some branches of the U.S. military continue to vaccinate against smallpox, owing to concerns that the virus could be used as a biological weapon.

ment awarded him £10,000 in 1802 and another £20,000 in 1807 as compensation for his valuable efforts on behalf of humanity.

Jenner was honored by Napoleon Bonaparte in France, and, in the United States, Thomas Jefferson requested vaccine from Jenner to vaccinate his family and neighbors at Monticello. In London, Jenner formed a society to promote the use of his vaccine to eradicate smallpox around the world; this organization became known as the National Vaccine Establishment in 1808. In 1821 Jenner was appointed "physician extraordinary" to King George IV.

TIME LINE

1749	1773	1796	1798	1808	1821	1823
Edward Jenner born in Gloucestershire, England.	Jenner completes his medical training.	Jenner conducts his first successful smallpox vaccination.	Jenner publishes *An Inquiry into the Causes and Effects of the Variolae Vaccinae.*	Jenner forms the National Vaccine Establishment.	Jenner appointed physician extraordinary to King George IV.	Jenner dies.

Despite the accolades, Jenner remained in his native Gloucestershire, corresponding about the progress of his vaccination around the world, collecting fossils, and propagating hybrid plants in his garden. In addition to his seminal work developing the cowpox vaccine, Jenner also made pioneering observations into heart disease. He performed several postmortem examinations on bodies of patients who had died complaining of chest pain—or angina pectoris. Jenner noted that the arteries around these patients' hearts had hardened and were blocked with fatty deposits. His observations advanced the practices of doctors treating heart disease. Jenner died of a stroke on January 26, 1823, at age 73.

In Jenner's native England, smallpox vaccination was made compulsory in 1853. Even so, smallpox outbreaks continued there into the 1960s, mainly after travelers returned from other countries where the disease still existed. In 1967, the World Health Organization (WHO) launched an aggressive worldwide campaign to end smallpox. That year the WHO estimated that some 15 million cases occurred each year, mainly in Africa, India, and South America. After 13 years of chasing down outbreaks in these areas, the WHO finally in 1980 declared "Smallpox Is Dead!" At the turn of the 21st century, the last remaining specimens of the disease were

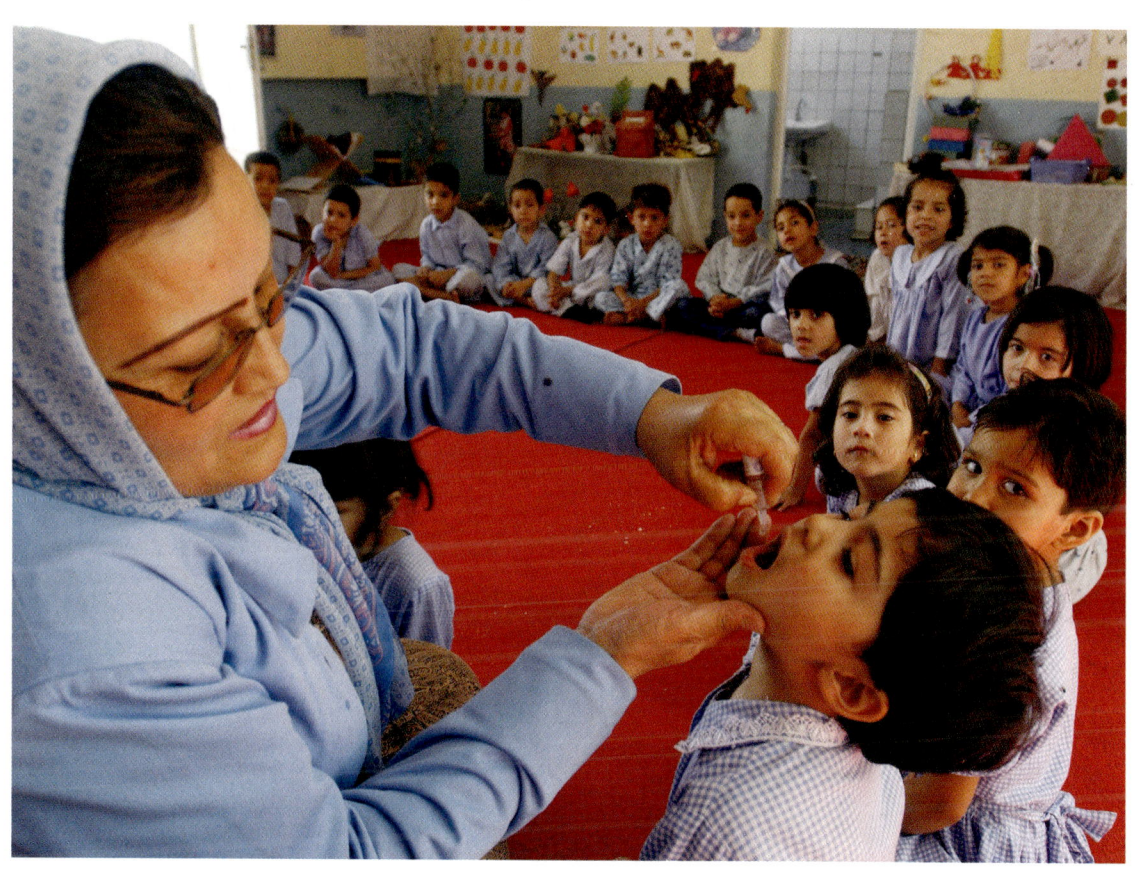

Vaccination remains an important weapon in the health care arsenal. In Kabul, Afghanistan, a health worker vaccinates school children against polio in 2005.

held under extremely close security in laboratories in the United States and Siberia.

Edward Jenner's pioneering work in immunology led to what many consider to be the most important branch of modern medicine. Called "the father of immunology," Jenner advanced our understanding of how foreign bodies—viruses, bacteria, chemicals—can interact with the human body and of how our bodies can be protected against them.

—Paul Schellinger

Further Reading

Books
Link, Kurt. *The Vaccine Controversy: The History, Use, and Safety of Vaccinations.* Westport, CT: Greenwood, 2005.

Rodriguez, Ana María. *Edward Jenner: Conqueror of Smallpox.* Berkeley Heights, NJ: Enslow, 2006.

Web site
Edward Jenner Museum
 Biography, history, and immunology resources.
 http://www.jennermuseum.com/

See also: Health and Medicine.

STEVE JOBS AND STEVE WOZNIAK

Inventors of the Apple I and Apple II computers
1955– and 1950–

Inventing usually begins with a great idea, but it does not end there. Convincing others that a new invention is something they need may be equally important. Many inventors dream that their ideas will improve the world; they try to sell not only a new gadget today, but also a vision of tomorrow. Two Californians, Steve Jobs and Steve Wozniak, used this approach to invention in the 1970s. When they developed their Apple computers, they sold a dream of how personal computers could change people's lives for the better.

EARLY YEARS

Jobs and Wozniak grew up in the same town, Los Altos, California, though Wozniak was five years the senior. Stephen "Woz" Wozniak was born August 11, 1950, and developed his lifelong passion for electronics by emulating his father, an engineer. Another big influence was Tom Swift, a fictional teenage hero in Victor Appleton's adventure novels, who used science and creativity to solve problems and save the world. By fifth grade, Wozniak had built his first electronics and computer projects. He excelled at math and science at Homestead High School and went on to study electronics and computer science at the University of California at Berkeley. In 1971, he quit his studies and was hired as an electronics engineer by the Hewlett-Packard computer company.

From left, Steve Jobs, John Sculley (then president of Apple), and Steve Wozniak pose with the Apple IIc (the compact version of the Apple II) 1984.

Steven Paul Jobs was born in Green Bay, Wisconsin, on February 24, 1955, and was raised by adoptive parents in California. Like Wozniak, he showed an early interest in electronics. Jobs was lucky enough to befriend a neighbor who built electronic appliances from kits, so he understood how they worked. After attending Homestead High in Los Altos, Jobs attended Reed College in Portland, Oregon, in 1972, but dropped out after only one semester. By 1974, he had been hired by the Atari Company to design computer games, but he left that job after one year to travel around India.

COMPUTER REVOLUTIONARIES

The early 1970s were a remarkable time to be involved in electronics. In 1971, an engineer working at the Intel Electronics Corporation, Ted Hoff (1937–), invented a way of putting all the essential components of a computer onto one tiny silicon chip. This invention, known as the microprocessor or microchip, made possible such devices as pocket calculators and digital watches. Before this, the smallest computers were called "minicomputers," but they were actually the size of a large washing machine—or bigger—and less powerful than the personal computers (PCs) in use today.

In 1975, Ed Roberts, an American electronics enthusiast, used the new microchip technology to launch the world's first computer in kit form, the Altair 8800. Very primitive, it had to be programmed by flipping switches on its case, and it showed the results of its operations using rows of tiny red lights. It was more like a piece of laboratory equipment than a modern computer. Nevertheless, the Altair 8800 proved to be a big hit with electronics hobbyists, causing a particular sensation in one Palo

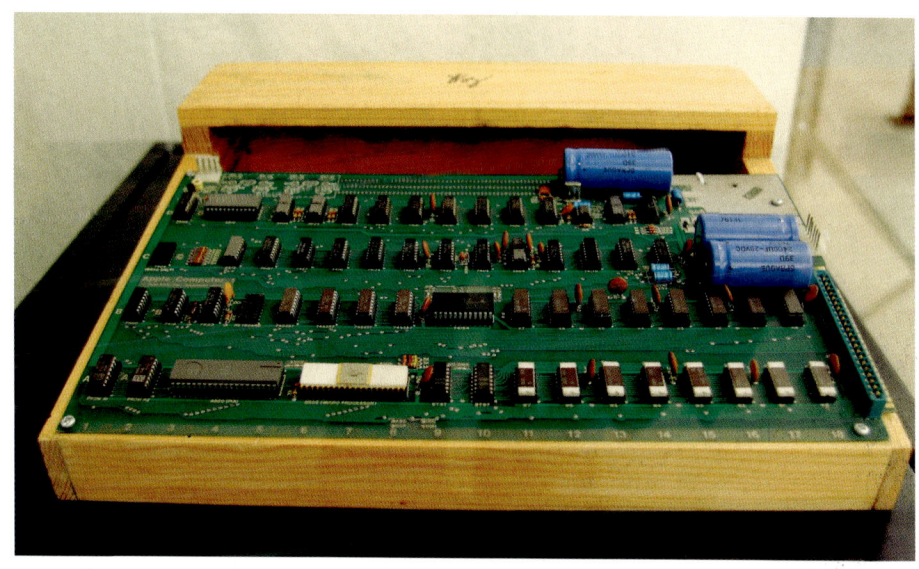

The Apple I computer, designed by Steve Wozniak in 1976; users would purchase the kit and add their own keyboards and screens.

Alto group calling themselves the Homebrew Computer Club. Among them were Steve Wozniak and his friend Steve Jobs.

Jobs and Wozniak first met at Hewlett-Packard, where Jobs had taken a summer job. By this time, Wozniak was already dabbling in the electronics business, making and selling small electronic gadgets called "blue boxes." Held near a telephone handset, a blue box would generate precise musical tones that could fool the exchange into allowing calls to be placed free. "Phone phreaking," as this was known, was popular in the early 1970s but was also illegal. Jobs and Wozniak sold the blue boxes for $150 each, although this "business" of theirs was run more for kicks and for undermining authority than to make money; for example, in the 1996 documentary *Triumph of the Nerds*, Wozniak gleefully recalls making a free call to the pope, while pretending to be Secretary of State Henry Kissinger.

> It wasn't like we both thought it was going to go a long way—it was like, we'll both do it for fun, and even though we're going to lose some money probably, we'll just have been able to say we had a company.
> —Steve Wozniak

Wozniak had often thought of building his own computer; seeing the Altair at the Homebrew Club convinced him he could do much better. He was proud to be a "hacker": among computer experts, a hacker is someone whose dogged determination and occasional flashes of inspiration permit him or her to solve difficult technical problems. Wozniak worked for months on the project and showed off the results to his fellow club members in 1976. Like the Altair, the computer he made, the Apple I, was based on a microprocessor and built on a single circuit board. Unlike the Altair, it was programmed by entering commands on a typewriter-style keyboard and it displayed results on an ordinary television set—it was altogether more "user-friendly." Members of the Homebrew Club were impressed, none more than Steve Jobs.

THE RISE OF APPLE COMPUTER

Wozniak has stated that he designed the Apple I just for the fun of doing it. However, when Jobs saw the machine, he saw the future. He persuaded Wozniak to go into business with him selling the Apple I in kit form. They needed money quickly, so Jobs sold his Volkswagen bus, Wozniak sold a valuable calculator, and together they raised $1,300 to start the business. On April 1, 1976, Wozniak and Jobs set up Apple Computer Corporation in the garage belonging to Jobs's parents. That year, they made and sold 175 kits priced at $666.66 each—far ahead of their expectations.

Wozniak was already working on a more advanced machine, the Apple II. It had more built-in memory, it had color graphics, and it could store programs using a cassette recorder. It could be programmed using the computer language BASIC—a series of English-like commands (such as

Steve Wozniak poses with an Apple II computer in 1986.

GOTO, INPUT, and PRINT) that even a novice could very quickly master. All these technical improvements were Wozniak's doing; however, the genius of the Apple II lay just as much in the way it was marketed—Steve Jobs's arena. He realized that the machine needed to be packaged in a nice-looking plastic case, like a television or a stereo, and it needed to appeal to ordinary families, not just computer hobbyists.

When the Apple II was launched in April 1977, it cost $1,298—twice as much as the Apple I. In the next few years, Apple Computer sold more than 50,000 of the machines. By the end of 1980, the little company Jobs and Wozniak formed in a garage had made its debut on the stock market, making its founders multimillionaires. The Apple II stayed in production for a decade and its success turned Apple into one of the world's biggest companies. Jobs and Wozniak were the toast of the computer world—but everything was about to change.

GOOD-BYE TO APPLE

Success brought Steve Wozniak a private plane, but when he crashed it in February 1981 he thought hard about his life and decided to leave Apple for a while. With more time for his personal life, he married and organized music and technology festivals. In 1983, he returned to the University of California at Berkeley to finish his degree, protecting his privacy by enrolling under the name Rocky "Raccoon" Clark. Later that year he returned to Apple to help develop new products, but he left for good in 1985. He has since devoted much of his time and money to educational and charity projects and teaches electronics and computing to young children

Steve Jobs also said good-bye to Apple in the 1980s, but much more dramatically than his partner. He had recruited John Sculley, a senior executive from the Pepsi soft drink company, to help him manage Apple's

Basic Elements of a Personal Computer

monitor

mouse

keyboard

Early personal computers such as the Apple I did not include many of the elements that are now considered standard, such a keyboard, monitor, and mouse.

1. Motherboard, the main circuitry of the computer to which all parts are connected.
2. Central processing unit (CPU), which processes instructions and data.
3. Random-access memory (RAM), the primary storage of the computer.
4. Expansion cards, which are small circuit boards used to control various components including sound and graphics.
5. Power supply.
6. Optical disk drive, which records data on disks using laser technology, such as CDs and DVDs.
7. Hard disk, a sealed unit that records saved data until the user erases or overwrites it.

Industrial Design

Apple pays a great deal of attention not just to the internals of its products (how they work and what they do), but also to their external appearance. Designing a product to appeal to its potential users is an essential part of modern inventing and is known as industrial design.

Apple's iPod music player is widely regarded as an excellent example. With an appealing and sophisticated look, the iPod has very few keys or switches to clutter its case, is easy to use, and has no lengthy instruction manual. Many people describe the design, created by Jonathan Ive (1967–), as "clean" and "timeless."

Industrial design helps to sell products by making them seem different from the competition. With electronic gadgets like MP3 players and cell phones, products from different manufacturers tend to work in similar ways and perform similar functions or actions; consumers may struggle to tell them apart or choose between them. For such products, industrial design can make all the difference between success and failure.

spectacular success. Jobs was considered to be a temperamental leader, with a reputation for bullying workers and pushing them to the limit; Sculley was a traditional corporate executive. Jobs and Sculley had very different ideas of the company they thought Apple should become, and Jobs was ousted in 1985 after a boardroom battle. Jobs later recalled, "What had been the focus of my entire adult life was gone, and it was devastating."

He soon developed new interests. One was a company called NeXT that made expensive computer workstations and software. Although technically advanced—Tim Berners-Lee used a NeXT computer to develop the World Wide Web—NeXT never achieved the success Jobs desired. His other project was Pixar, a graphics studio that developed computer technology for the movies. From the mid-1990s onward, Pixar produced a series of spectacularly successful animated movies, including *Toy Story*, *A Bug's Life*, and *Monsters, Inc.*, and made Jobs a billionaire.

THE RETURN OF STEVE JOBS

By this time, Apple was in deep trouble. The Macintosh was still using 1980s technology, and its archrivals, Microsoft and Intel, had joined forces to dominate the personal-computer business. Apple's board fired Sculley and hired other managers to halt the decline, but the situation grew steadily worse. In 1996, Steve Jobs agreed to return to Apple as an unpaid adviser; the following year, he was back in charge. Paid a salary of just $1 per year, Jobs was listed in the *Guinness Book of Records* as "The World's Lowest-Paid Chief Executive Officer."

Jobs had lost none of his marketing flair. In the 1970s, he had used his charisma to launch the Apple II as a "cool" product that every computer enthusiast should aspire to own. His amazing knack for selling products led one Apple engineer to suggest that Jobs generated a "reality distortion field (RDF)," a way of mesmerizing employees into accomplishing the impossible and charming customers into buying things they did not really need. In 1984, Jobs repeated that trick with Apple's Macintosh. After returning to Apple in the 1990s, he worked the same magic with the iMac (1998), a redesigned version of the Macintosh; the iPod portable music and video player (2001); and the online music store iTunes (2003). Taking Apple back to its roots, with innovative products and attractive design, Jobs soon restored the company's vision—and its fortune. Nevertheless, questions continue about the close link between Jobs and Apple. Is the company overly dependent on him? In 2004, when news broke that Jobs had had cancer surgery, Apple's stock price dipped significantly.

THE IMPORTANCE OF APPLE

With the Apple I and Apple II, Steve Jobs and Steve Wozniak invented not only a pair of technically brilliant personal computers but also the concept of user-friendly personal computing. Until the Apple II went on sale in 1977, computing was just a hobby for electronics enthusiasts.

TIME LINE

1950	1955	1971	1974	1976	1977
Stephen Wozniak born in Los Altos, California.	Steven Paul Jobs born in Green Bay, Wisconsin.	Wozniak hired by Hewlett-Packard.	Jobs hired by the Atari Company.	Jobs and Wozniak found Apple Computer.	Wozniak and Jobs launch the Apple II.

Jobs with Bono and The Edge of U2 at an iPod event in 2004. Apple cultivates a "cool" image and many celebrities want to be associated with the company.

When the Apple II became the computer of choice for small businesses in the late 1970s, the world's biggest computer company, IBM, was forced to launch its own personal computer, and it did in 1981. These actions resulted in most personal computers working the same way, running compatible software, and becoming easier to use. The personal-computer revolution was under way.

Jobs and Wozniak had envisioned this revolution many years earlier. As Jobs said later: "The thing that bound us together at Apple was the ability to make things that were going to change the world." Apple's prod-

TIME LINE

1981	1984	1985	1980s	1996	2001
Wozniak leaves Apple.	Apple releases the Macintosh.	Jobs forced to leave Apple.	Jobs founds NeXT.	Jobs returns to Apple.	Apple introduces the iPod.

> Your work is going to fill a large part of your life, and the only way to be truly satisfied is to do what you believe is great work. Don't settle.
> —Steve Jobs

ucts are still personally launched by Steve Jobs and are marketed in exactly the same way as was the Apple II—they claim to be more innovative, easier to use, and better-designed than the competition (see box, Industrial Design).

Apple products have always been known for their combination of technical innovation and appealing design—this is the legacy of Wozniak and Jobs. Jobs provided the vision; Wozniak (and his successors at Apple) made technical breakthroughs that enabled the vision to become a reality. The Apple story suggests that modern inventing involves marketing as much as innovation. Although neither technical brilliance nor marketing alone is enough to guarantee that an invention will succeed, together they often make a winning combination.

—Chris Woodford

Further Reading

Books
Deutschmann, Alan. *The Second Coming of Steve Jobs*. New York: Broadway, 2001.
Kendall, Martha. *Steve Wozniak: Inventor of the Apple Computer*. Los Gatos, CA: Highland, 2000.
Levy, Steven. *Hackers: Heroes of the Computer Revolution*. New York: Penguin, 2001.

Web sites
Apple Computer
    Official Web site of Steve Jobs's Apple Computer Corporation.
    http://www.apple.com/
Steve Wozniak
    Personal Web site of Steve Wozniak, with links to historical Apple Web sites.
    http://woz.org/

See also: Berners-Lee, Tim; Computers; Entertainment; Kilby, Jack, and Robert Noyce.

LONNIE JOHNSON

Inventor of the Super Soaker

1949–

Lonnie Johnson is an inventor who refused to specialize. His company, Johnson Research & Development, develops technology for toys and for more serious applications, such as fuel cells. Johnson developed the Super Soaker, a high-powered water gun that remains a popular toy, while working at the National Aeronautics and Space Administration's Jet Propulsion Laboratory.

EARLY YEARS

Lonnie Johnson was born in 1949, the third of six children. He grew up in Mobile, Georgia, where his father worked as a truck driver for the U.S. Air Force, and where his mother worked as a nurse's aide. Johnson showed an interest in machinery early, frequently taking his siblings' toys apart to see how they worked.

Johnson also became interested in rocketry at a young age. This interest led to some dramatic incidents. One time he was mixing rocket fuel in the kitchen when the fuel started a fire, which was put out without any serious injury. When Johnson was 14, he took some of his rocket fuel to school. A classmate set the fuel on fire in a hallway, and Johnson was taken to the police station and accused of trying to blow up the school.

Still, Johnson did well in high school. In 1968 he built a remote-controlled robot that won a national engineering competition. He won a scholarship to Tuskegee University, where he earned a bachelor of science in 1972 and a master's degree in nuclear engineering in 1974.

DEVELOPING THE SUPER SOAKER

A few months after receiving his master's degree, Johnson joined the air force. While in the service, he worked in engineering positions in a variety of fields, ranging from nuclear power to computing. In the early 1980s he was transferred to the Jet Propulsion Laboratory (JPL) in Pasadena, California, where he worked on developing satellites and probes for various space missions, including the Galileo probe of Jupiter and the Mars Observer project.

Johnson poses with his invention outside his home in Atlanta, Georgia, in 1992.

Johnson enjoyed his work, but he became frustrated with the bureaucracy and restrictions of the air force and JPL. He tinkered at home, working on various

Using Air to Shoot Water

Before the advent of the Super Soaker, water guns had a simple design. Water was held in a reservoir chamber, usually located in the handle of the gun. A second chamber was connected to the first through a one-way valve, which was connected to the nozzle.

The gun's trigger acted as a pump. When the trigger was released, it would suck water from the reservoir into the second chamber, and the one-way valve would hold the water there. When the trigger was squeezed, the water would be expelled from the second chamber out the nozzle of the gun. As the squeezing of the trigger was the only force that expelled the water out of the gun, the gun did not shoot much water and did not shoot it very far.

The Super Soaker, in contrast, eliminated the pumping action of the conventional water-gun trigger. Instead, the gun had two reservoir chambers, one of which was airtight. Before shooting the gun, the user would pump water from one reservoir chamber into the second, airtight one, to ready the gun to fire. As a result of this pumping action, the air within the second chamber would become compressed. Instead of acting as a pump, the trigger opened a connection between the second chamber and the nozzle. Once that connection was open, the compressed air forced the water out of the gun at high pressure.

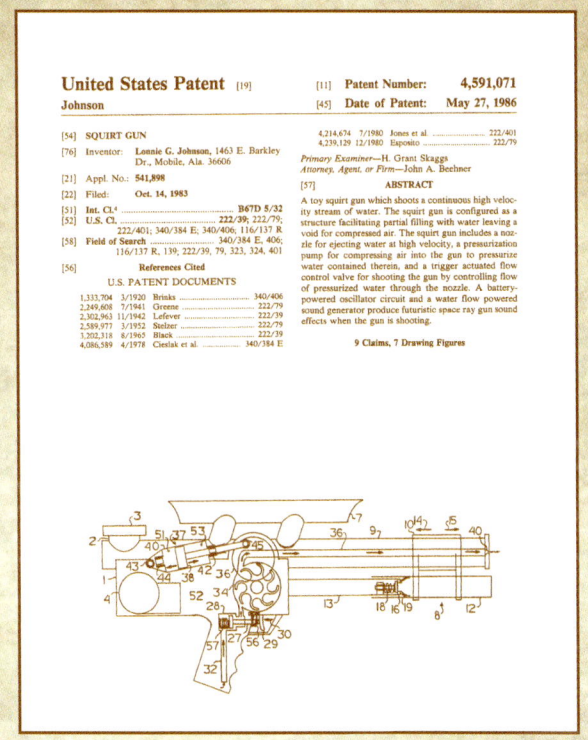

Johnson's 1983 patent on a high-velocity squirt gun that would become the Super Soaker.

inventions of his own. One day in 1982, Johnson was in his bathroom working on a new kind of water pump. He attached a high-pressure nozzle to the sink and was surprised when the water shot all the way across the bathroom.

Johnson thought the nozzle might be used to make a water gun for his nine-year-old daughter, so he created a prototype for her. The powerful water gun proved a huge hit in the neighborhood, and Johnson realized that he might have a marketable product.

FALSE START

Johnson filed for a patent for the new water gun the next year. It would take many more years for the Super Soaker to become a reality, however. Johnson was not a manufacturer, and he did not have the money to build and market the gun himself. He spent the next several years trying to interest toy companies and investment banks in the water gun and his many other inventions.

In 1987 an investment firm agreed to finance and develop some of his ideas. Johnson, assuming that the deal would go through and that he would have a new career, quit the air force and JPL, and arranged to sell his home. At the last minute, however, the firm reneged on the agreement. Johnson was able to get his job with JPL back, but when he backed out of his agreement to sell his house, he was sued.

ON THE MARKET

Johnson continued his efforts to sell his water gun. Finally, during a trade show in New York City, an executive from Larami Corporation, a small toy company, saw the gun and liked it. Larami reached a licensing agreement with Johnson in 1989, and his gun, first called the Power Drencher and then the Super Soaker, went on the market at last.

The Super Soaker was far more powerful than the standard water gun. An ordinary water gun had a range of just a few feet, but a Super

TIME LINE

1949	1968	1974	1982	1989	1995
Lonnie Johnson born; he grows up in Mobile, Alabama.	Johnson wins a national engineering competition.	Johnson earns a master's degree in nuclear engineering.	Johnson creates a high-velocity squirt gun.	The Super Soaker goes on the market.	The Super Soaker line of products is sold to Hasbro, Inc.

Playing with contemporary versions of the Super Soaker.

Soaker could hit a target 40 feet (12 m) away. Sales for the gun took off in 1991 after Johnny Carson, host of *The Tonight Show*, used a Super Soaker to blast his staff during the program.

By 1992, sales of Super Soakers topped $200 million. The success of the gun gave rise to a slew of improvements and modifications, resulting in an entire line of Super Soaker water guns, as well as an untold number of imitations. Three years later, Larami sold the Super Soaker line to Hasbro, Inc., the second-largest toy maker in the United States.

The Super Soaker brand has continued to sell well and accounts for roughly 85 percent of the water-gun market.

The success of the Super Soaker allowed Johnson to resign from the JPL and to become a full-time inventor. He currently runs Johnson Research & Development in Atlanta, Georgia, a firm that has developed toys such as Nerf guns and air-powered rockets. In addition to toys, Johnson has received research contracts from various U.S. government agencies to develop devices ranging from rechargeable batteries made of thin film to environmentally safe refrigeration systems and advanced fuel cells.

—Mary Sisson

> There is no short easy route to success. . . . It takes a lot of hard work and a bit of luck to be successful.
> —Lonnie Johnson

Further Reading

Book
Amram, Fred. *African-American Inventors*. Mankato, MN: Capstone, 1996.

Web sites
Johnson Research & Development
    Web site of Johnson's company.
    http://www.johnsonrd.com
Super Soaker
    Official Web site for the Super Soaker.
    http://www.hasbro.com/supersoaker/

See also: Entertainment.

AMANDA JONES

Inventor of vacuum canning
1835–1914

Amanda Jones created a low-heat method of canning food. She was not, however, only an inventor—she firmly believed that as a psychic medium, she received instructions from the spirits of the deceased. She founded the Woman's Canning and Preserving Company in 1890 to manufacture food that had been canned using her technique.

EARLY YEARS

Amanda Theodosia Jones was born in 1835 in East Bloomfield, New York, to a large family. Her father was a weaver, and the family was not wealthy. Both of her parents were avid readers who valued education.

In 1845 Jones's family moved to a small town near Buffalo, New York. One day, Jones was attending school with one of her brothers when he suddenly collapsed and died. That experience elicited a fascination with spiritualism in Jones, who wanted to contact her dead brother.

Spiritualism, a religious movement that became popular in the 1840s, claimed that the spirits of the deceased could be contacted by gifted individuals known as mediums. Mediums could allegedly communicate with spirits while asleep or in a trance state; often a medium claimed to have a particular spirit guide who would speak through him or her, providing advice and insight.

In 1850, Jones graduated from a normal school, where teachers were trained, and she began teaching. She also began writing poetry; her first poems were published in 1854. Jones then quit teaching to focus on her writing.

Jones realized that foods could be canned at a low temperature if the air was removed or replaced with juice or syrup.

VACUUM CANNING

During the 1860s, Jones continued to publish poems and worked as an editor. She had chronic health problems, and her interest in spiritualism intensified, even leading her to believe that she herself was a medium. She eventually claimed to have been in contact with various spirit guides who would instruct her to move or take up new occupations.

In 1872, Jones had an idea for a new way to can fruit. Canning had been invented in the early 1800s as a means of preserving food for long periods. At the time, canning involved first sealing food in a can or jar, then cooking it at high temperature for a long time. As a result, canning food was grueling, and canned food was not nearly as flavorful or nutritious as fresh food.

Jones would later claim that her spirit guides suggested that food could be canned without cooking it. Jones got the idea that food could perhaps be canned at a lower temperature if all the air was vacuumed out of the can and replaced with fruit juice or syrup.

Jones traveled to the home of a distant relative, LeRoy Cooley, living in Albany, New York. Jones and Cooley went to work developing Jones's canning idea, and in 1873 they were granted several patents for what would become known as the Jones process.

Fruit to be canned by the Jones process was placed in a jar. An air pump was then used to remove as much air as possible from the jar. Juice or syrup was then poured into the jar, which was subsequently heated, causing the fluids inside to boil. Because the air pressure was so low in the jar, boiling would occur at temperatures as low as 100 degrees Fahrenheit (37.8°C). Low temperatures created an additional complication: they were not high enough to kill bacteria in food, which traditional canning,

> This is a woman's industry. No man will vote our stock, transact our business, pronounce on women's wages, supervise our factories. Give men whatever work is suitable, but keep the governing power.
>
> —Amanda Jones, addressing employees of the Woman's Canning and Preserving Company

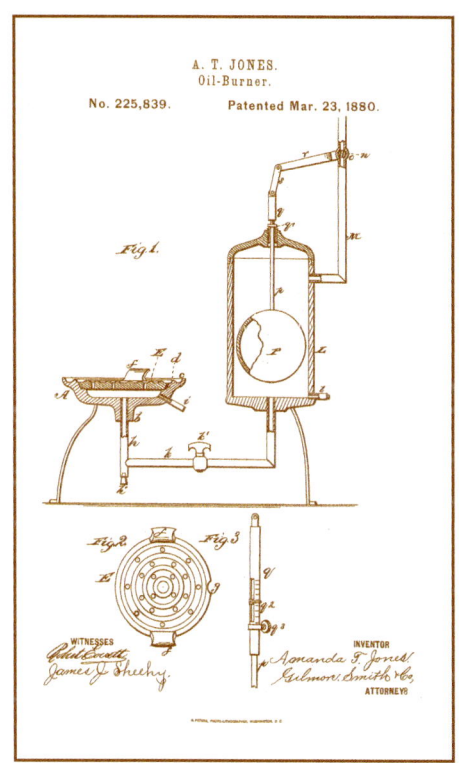

Illustrations from Jones's 1880 patent for automatic safety burner.

with its high temperatures, did. The process remained effective nonetheless because it deprived bacteria and molds of oxygen, which they needed to survive and grow.

Although Jones and Cooley patented their ideas, they did not attempt to turn them into a business at the time. Instead, Jones moved on to other interests.

BURNING OIL

In 1880 Jones traveled to western Pennsylvania to examine that area's oil fields. At the time, this form of oil was a relatively new fuel. People were accustomed to burning solid fuel such as coal; liquid oil tended to spill and cause fires.

Jones developed and patented an automatic safety burner, which had a valve to control the amount of oil released into the burner. Tests of the new burner helped Jones to attract a financial backer; however, the backer's fortune was lost in a stock-market collapse, leaving Jones without the funds she needed to turn her idea into a business.

THE WOMAN'S CANNING AND PRESERVING COMPANY

In 1890 Jones returned to her vacuum-canning idea, founding a company, the Woman's Canning and Preserving Company, in Chicago. As the company's name suggested, all of the employees were women (except Cooley and the man who operated the building's boiler), as were all of its stockholders.

The company began selling canned rice and tapioca puddings, eventually moving into canned meat. Woman's Canning expanded, opening two new facilities. The company's other executives began to lose confidence in Jones's management, however; she was forced out of the business in 1893.

TIME LINE

1835	1850	1872	1880	1890	1893	1914
Amanda Jones born in East Bloomfield, New York.	Jones graduates from normal school.	Jones develops a new canning method.	Jones develops an automatic safety burner.	Jones founds the Woman's Canning and Preserving Company.	Jones is forced out of her company.	Jones dies.

The Spirits of Invention

Jones's desire to be recognized as a psychic medium and inventor created a unique conundrum about assigning credit for her ideas: her alleged contact with the spirit world and her insistence that she dutifully followed the instructions of spirits could suggest that she was using the ideas of others. The issue was particularly problematic for Jones because her spirit guides were usually male. At the time, men were considered by many to be intellectually superior to women.

Sensitive to the possibility of being regarded as intellectually inferior, Jones detailed in her autobiography precisely how her spirit guides assisted her: they told her about patent law, that it was possible to can fruit without cooking it, and that she should work with Cooley to develop the idea.

She insisted, however, that the Jones process was her own invention, created without explicit instruction by spirits. She maintained that she woke up one day with the basic idea that later she and Cooley developed into a workable process without the benefit of otherworldly instruction. "Now, let me say at once, no spirit told me this," she wrote. "I have inventions—patentable—patented. They are as much my own as are my many poems—mostly studied out by slow and painful process, often at bitter cost. To every patent application I have taken oath, unperjured: "This is my invention.—This I claim."

LATER YEARS

Undaunted by the loss of control over her company, Jones continued to patent improvements to her burner and the Jones process. She also continued to write, publishing her autobiography in 1910. She died in 1914, seven years before Woman's Canning went out of business.

Vacuum canning remains a standard method for foods that will lose flavor if exposed to high heat. The process has been modified over the years for use with dry foodstuffs like coffee, which is vacuum-canned and packed with nitrogen rather than juice or syrup. Jones's innovations made canning a versatile method of preserving food for longer periods with less loss of flavor and nutrition.

—Mary Sisson

Vacuum canning remains a popular method of food storage. A jar of unopened pickles can last for as long as 12 months.

Further Reading

Books
Macdonald, Anne L. *Feminine Ingenuity: Women and Invention in America*. New York: Ballantine, 1992.

Vare, Ethlie Ann, and Greg Ptacek. *Mothers of Invention*. New York: Morrow, 1988.

Web sites
How Do I? . . . Can
> Information on canning from the National Center for Home Food Preservation.
> http://www.uga.edu/nchfp/how/general.html

Spiritualism
> An introduction to the religious movement by the University of Virginia.
> http://religiousmovements.lib.virginia.edu/nrms/Spiritsm.html

See also: Appert, Nicolas-François; Food and Agriculture.

PERCY LAVON JULIAN

Developer of a treatment for glaucoma and a method of synthesizing cortisone
1899–1975

To be able to create, all inventors must overcome barriers. Percy Lavon Julian faced exclusion from schools, jobs, and neighborhoods because he was African American. Nevertheless, Julian made important contributions to science and to medicine, developing methods to cheaply synthesize drugs that have helped countless people.

EARLY YEARS

Percy Lavon Julian was born in Montgomery, Alabama, on April 11, 1899, the oldest of six children. His grandparents had been slaves, and his father worked as a railway mail clerk.

Julian's parents strongly encouraged their children to become educated, but laws in Alabama forced Julian to attend segregated elementary schools that offered inferior education. Moreover, in Montgomery, no high school accepted black students. Instead of attending high school, Julian attended a trade school.

Although the schools Julian attended offered few science classes, he was fascinated by science and read about it on his own. He particularly liked chemistry, an interest that troubled his parents, who were concerned that racial discrimination would make it impossible for an African American to make a living as a chemist.

Julian persuaded them to let him try, however, and in 1916 he gained admission to DePauw University in Greencastle, Indiana. Since Julian had not attended an academic high school, he was required to take classes at the high-school level in addition to regular college classes. He also had to work to finance his tuition. Despite these challenges, Julian graduated from DePauw in 1920 as class valedictorian, an achievement that so delighted his parents that they moved the family to Greencastle and sent all of Julian's siblings to DePauw.

THE STRUGGLE TO BE A BLACK ACADEMIC

Given his strong academic record at DePauw, Julian expected to be awarded a fellowship that would allow him to continue studying chemistry as a graduate student. Such fellowships were not forthcoming because of his race, however, so Julian took a job teaching chemistry at Fisk University, a black college in Nashville, Tennessee.

Julian's professors at DePauw kept lobbying for him. Finally in 1922 Julian was awarded a fellowship to Harvard University in Cambridge, Massachusetts, where he studied organic chemistry—the chemistry of carbon compounds, the basis of living things. Julian received his master's degree in 1923 and continued to work at Harvard as a researcher for a couple of years, hoping for a permanent position.

The position did not materialize, however, so Julian left Harvard and went back to teaching, becoming the only professor of chemistry at West Virginia Collegiate Institute. In 1928, he moved to Washington, D.C., to teach at Howard University, a black college.

A year later, Julian received a fellowship to study for a doctorate in organic chemistry at the University of Vienna in Austria. The opportunity was the realization of a dream and offered a chance to study under

Ernst Späth (1886–1946), an extremely well-respected chemist who had been synthesizing naturally occurring chemicals in his laboratory.

Percy Julian after receiving an award in the 1950s.

GLIDDEN

Julian received his doctorate in 1931 and returned to Howard. A dispute with the administration led him to leave Howard a year later and take a research position at DePauw. There, he and a colleague developed a method to synthesize a drug to treat glaucoma, an eye disease (see box, A Treatment for Glaucoma).

Despite Julian's success, his race still stood in the way of his professional advancement. The Institute of Paper Chemistry wanted to hire Julian, but it was located in the town of Appleton, Wisconsin, where a law forbade African Americans from staying overnight.

One day the Institute's board was discussing the impossibility of hiring Julian, which was a source of great frustration to the executives. One member of the board was also an executive at Glidden Company, a paint and chemical company located in Chicago. Glidden used soybeans to make some of its products and wanted to employ an organic chemist to help develop more lucrative products. As Chicago had no laws forbidding African Americans from residing there, the executive called Julian and offered him a job at Glidden.

Julian had wanted to be a professor at DePauw, but the university would not appoint an African American to a professorship, so he accepted the offer from Glidden. In 1936, he went to work for the company as director of research for its soybean division.

SOYBEANS

At the time, soybeans were not widely used in the United States. Julian had studied the soybean while in Vienna and knew that the bean contained a plethora of compounds. At Glidden, he developed methods to purify useful products from soybeans, including lecithin, used to keep processed foods from separating; and soybean meal, an important source of protein in animal feed. He also developed a fire-suppressant foam made from soybean protein that was widely used by the U.S. Navy during World War II to fight fires on ships.

What really interested Julian about soybeans was their chemical compounds, which are very similar to human hormones. Julian thought it might be possible to make human hormones from soy compounds, but the process used to extract the compounds from soy oil destroyed the valuable oil.

A break came one day in 1940, when water leaked into a tank containing soy oil, creating a white mass. Julian tested the mass and discovered that it contained large amounts of the hormone-like soy com-

A Treatment for Glaucoma

At DePauw in the 1930s, Julian and a colleague he had worked with in Vienna, Joseph Pikl, began working on synthesizing the organic chemical physostigmine, which occurs naturally in small quantities in the West African Calabar bean. Physostigmine had been shown to be an effective treatment for some forms of glaucoma, a disease in which pressure builds up in the eye, eventually causing blindness.

Extracting enough physostigmine from Calabar beans to create a useful dose of medicine was an expensive and difficult process. Julian and Pikl went to work trying to synthesize physostigmine more cheaply in the laboratory.

They were not alone in trying to synthesize the chemical: a group of British scientists led by the respected chemist Robert Robinson, who would go on to win the Nobel Prize, had also been trying to make physostigmine in the laboratory. Robinson's team thought they were very close to a solution, and most scientists assumed that Robinson would solve the physostigmine problem.

In 1935, however, Julian and Pikl published a paper showing that they had successfully created physostigmine in the laboratory. Not only had they beaten Robinson, but their paper also proved that Robinson's work was deeply in error. The reversal astounded the scientific world and gave glaucoma patients an inexpensive and reliable new source of medication to help save their sight.

The soybean plant contains chemicals that are similar to human hormones.

pounds. The oil had not been ruined, so Julian developed a separation process based on what had happened during the leak. The new process allowed him to obtain the soy compounds in large quantities. He then developed methods to transform the hormone-like compounds into actual human hormones. He soon created a method to cheaply synthesize the human hormone progesterone, which is used by doctors to help pregnant women avoid miscarriage.

CORTISONE

Julian moved on to develop ways to synthesize other hormones from soy. In 1948, scientists discovered that the hormone cortisone could be an effective treatment for rheumatoid arthritis; Julian quickly began work on synthesizing cortisone.

Rheumatoid arthritis is a crippling ailment that develops when the body's immune system attacks its own joints. Cortisone suppresses the immune system, providing relief to sufferers of the disease. In addition, cortisone was also effective in treating other illnesses caused by immune system malfunction. In its natural form, however, cortisone was very expensive, costing about $700 per gram, and most of those who needed it could not afford it. In late 1949, Julian developed a method to create, from the soybean synthetic cortisone, that cost only about 50 cents per gram.

TIME LINE

1899	1920	1931	1936	1949	1954	1973	1975
Percy Julian born in Montgomery, Alabama.	Julian graduates from DePauw University.	Julian earns doctorate in organic chemistry.	Julian accepts position at Glidden Company.	Julian develops method for creating synthetic cortisone.	Julian founds Julian Laboratories.	Julian elected to the National Academy of Science.	Julian dies.

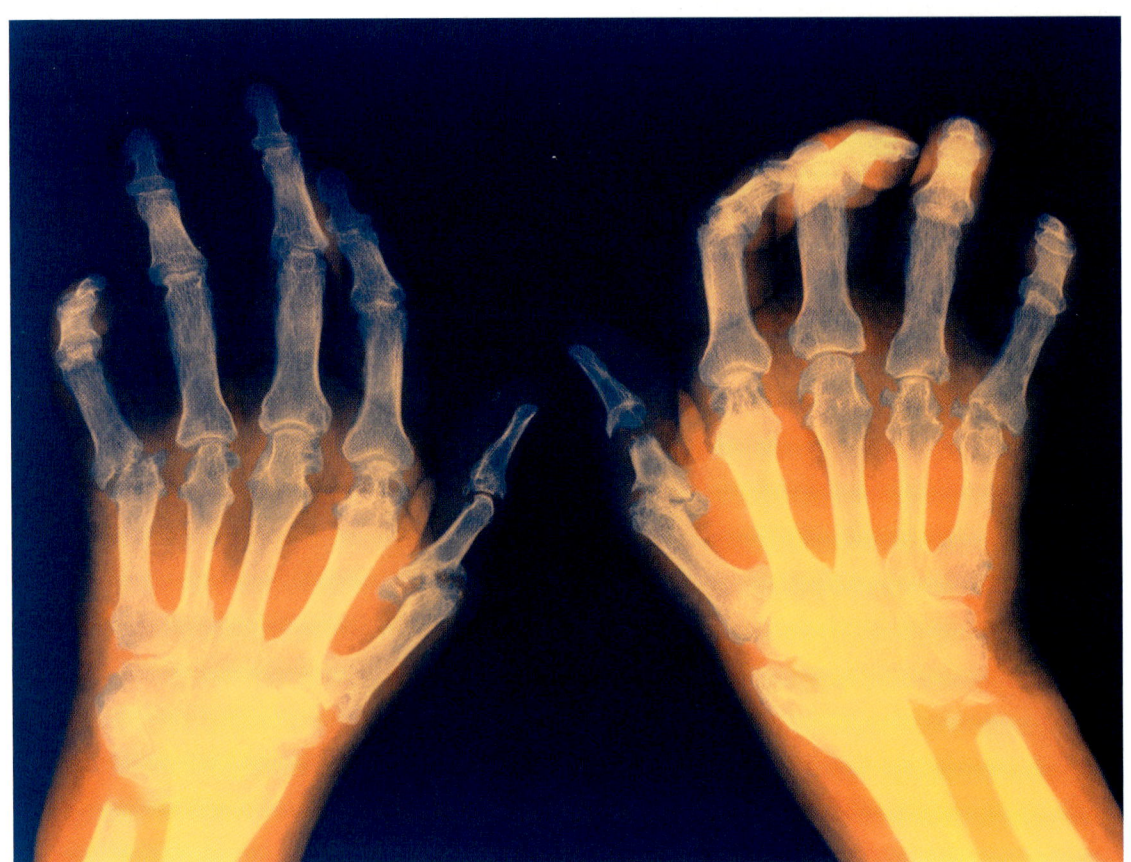

A color-enhanced x-ray shows the impact of rheumatoid arthritis, which can be treated with cortisone.

Julian's work provided much-needed relief to sufferers of rheumatoid arthritis and brought Julian renown. At times, however, his elevated public profile made him a target for racial violence. In 1950, Julian and his family moved into the wealthy, largely white Chicago suburb of Oak Park. Before the family moved in, vandals tried to set fire to the house, and several months later the house was bombed.

LATER YEARS

In 1953 Julian resigned from Glidden because he wanted to extend his research from soybeans to other promising plants. The following year he founded Julian Laboratories in Oak Park and began studying a wild yam found in Mexico that contained many promising hormone-like compounds. Julian developed methods to synthesize cortisone and other useful chemicals from the yam and built factories in Mexico and Guatemala to process them.

In 1961 Julian sold his business to the large pharmaceutical firm Smith, Kline, and French, his first step toward retirement. In later life, Julian was involved in a variety of nonprofit work in science education

Percy Julian at work in the lab around 1958.

and civil rights. He was elected to the National Academy of Science in 1973; two years later, he died at the age of 76.

Julian had a major impact on the world of drug development. His work created drugs that prevented people from going blind and becoming crippled; and he introduced new techniques for identifying, purifying, and synthesizing useful organic compounds.

—Mary Sisson

> My dear friends, who daily climb uncertain hills in the countries of their minds, hills that have to do with the future of our country and of our children, may I humbly submit to you, the only thing that has enabled me to keep doing the creative work, was the constant determination: Take heart! Go farther on!
>
> —Percy Lavon Julian

Further Reading

Book

Jenkins, Edward S., et al. *American Black Scientists and Inventors.* Washington, D.C.: National Science Teachers Association, 1975.

Web sites

Hall of Fame/Inventor Profile
  A sketch of Julian from the National Inventors Hall of Fame Foundation.
  http://www.invent.org/hall_of_fame/84.html

The Life and Science of Percy Julian
  A Science Alive! exhibition on the inventor.
  http://www.chemheritage.org/scialive/julian/index.html

Percy Lavon Julian
  A memoir of Julian from the National Academies Press.
  http://newton.nap.edu/html/biomems/pjulian.html

See also: Accidents and Mistakes; Bath, Patricia; Drew, Charles; Health and Medicine.

DEAN KAMEN

Inventor of the AutoSyringe and
the Segway

1951–

Dean Kamen spent much of his life developing lifesaving medical devices in relative obscurity. Kamen achieved celebrity with a mystery invention, ultimately revealed to be the Segway scooter. Ironically, the invention that brought Kamen fame had less of an impact than his earlier work in the medical field.

EARLY YEARS

Dean Kamen was born in 1951 and grew up in Rockville Centre, a town located on Long Island, New York. Kamen's father was a successful illustrator and his mother was a teacher; both parents wanted their children to do well in school. His older brother Bart was an exceptional student and eventually became a physician. Dean, in contrast, found school uninteresting and restrictive. He became a discipline problem and deliberately got poor grades on tests.

During high school, Kamen began tinkering with machinery on his own. He began designing light boxes, used to create audiovisual presentations. By the time he graduated from high school, Kamen was operating a successful small business out of his parents' basement.

At his parents' insistence, Kamen went to college, attending the Worcester Polytechnic Institute in Massachusetts. Although he attended classes, he devoted much of his time (he often went home on the weekends) to his light-box business.

MEDICAL DEVICES

While Dean Kamen attended Worcester Polytechnic, Bart Kamen attended Harvard Medical School. Bart noticed that the intravenous units used in hospitals to provide patients with fluids, nutrients, and drugs were difficult to use. The big, bulky units needed constant monitoring to ensure that they were providing patients with the correct drug dosage.

Dean Kamen in 2002.

As a result of inventions like the AutoSyringe, people in need of kidney dialysis are now able to perform the procedure at home.

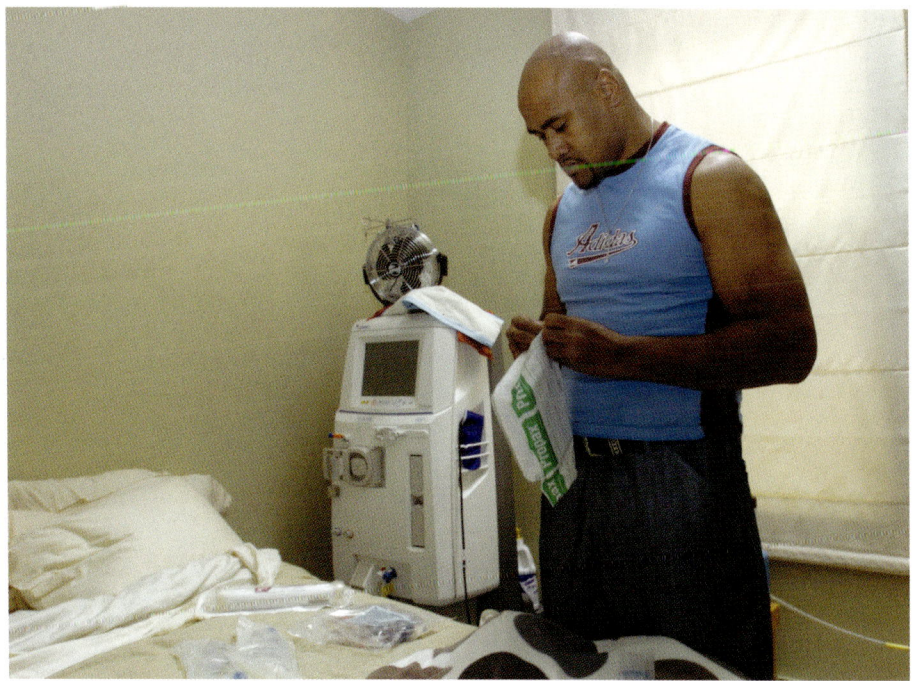

Bart suggested designing a new, automatic intravenous device to his younger brother. Intrigued by the challenge, Kamen went to work, building what would become the first drug infusion pump in 1971, when he was 20 years old. Infusion pumps administer medications directly into patients' circulatory systems. The prestigious *New England Journal of Medicine* did a story on the pump, and Kamen was flooded with inquiries.

Work on the pump took up so much of Kamen's time that he dropped out of college. In 1976, he started a new company, AutoSyringe, Inc., and began manufacturing the pumps. Frustrated by New York's high taxes, Kamen moved AutoSyringe to New Hampshire in 1979. AutoSyringe continued to grow, and Kamen eventually decided that he did not want to run a large manufacturing company. In 1982 he sold the company to the large drug and medical-device company Baxter International.

DEKA RESEARCH AND DEVELOPMENT

With the money he received from the sale of AutoSyringe, Kamen decided to start his own research and development company. Kamen combined the first two letters of his first and last name to form the company's name, DEKA Research and Development Corp. DEKA was located in Manchester, New Hampshire, where it created new devices and sold the rights to manufacture them to other companies.

In 1987, Baxter asked DEKA to improve its dialysis machine. A dialysis machine cleans waste products from the blood so that they do not poison the body, a function normally performed by the kidneys.

Existing dialysis machines were large, heavy, noisy, expensive, and difficult to operate and maintain. As a result, the machines were usually located only in hospitals or clinics, and could not be used at home by patients suffering from kidney failure. Additionally, some patients needed dialysis as often as twice a week and the process took several hours.

Kamen and his staff dramatically redesigned the dialysis machine. Their machine, which went on the market in 1993, weighed less than 25 pounds, was relatively cheap, and was fairly simple to operate. Thus individuals could run it at home. The machine was also very quiet—patients were able to undergo dialysis while they slept. The new machine made life far easier for people suffering from kidney failure—some could work and even travel.

A BETTER WHEELCHAIR

One day in the late 1980s, Kamen was walking down the street when he saw a man in a wheelchair struggling to get the chair over a curb. The then man went into an ice cream store, where he had to struggle again to reach across the counter to make his purchase.

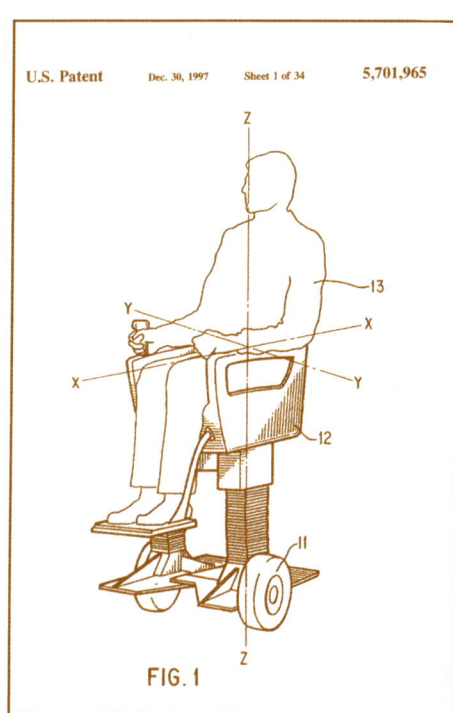
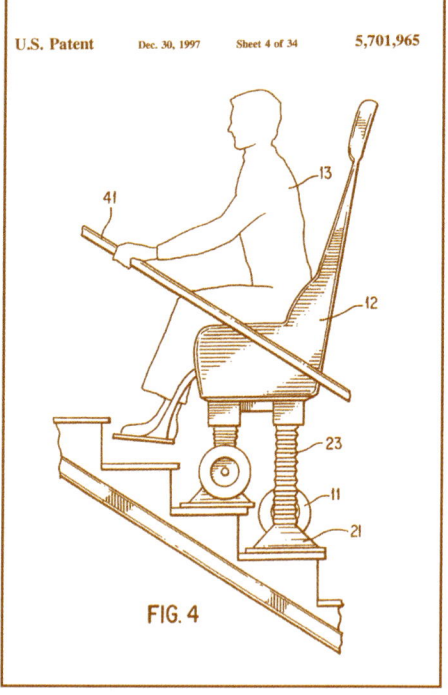

Illustrations from Kamen's 1997 patent for an improved wheelchair or "human transporter," designed to transport a person on irregular surfaces such as stairs.

For Inspiration and Recognition of Science and Technology

Kamen's own experience in school was far from ideal—despite his strong interest in science and engineering, he found his science and math classes dull. As he grew older, Kamen began to see his experience as symptomatic of a larger problem: although science can be very exciting, it is often not presented as such to young people. To Kamen, this attitude toward science is wrongheaded and potentially very harmful. He knows that the vast majority of people will never be talented and lucky enough to become stars in sports, movies, or music, but that many of them could be productive scientists. Kamen argues that the economic future of the United States depends on technological advancement, and that because of the current lack of glamour surrounding science, fewer young people go into the field, and so there are fewer inventions and advances.

To help young people see the value of pursuing science, Kamen created a nonprofit organization, FIRST (For Inspiration and Recognition of Science and Technology) in 1989. FIRST hosted a robotics competition in which teams of high school students designed and built robots to perform certain tasks, which changed from year to year.

FIRST has expanded greatly since its inception and has added a competition using robots made of Legos for younger students. By 2006 FIRST was attracting ten thousand students from around the world to its competitions, which have appeared on national television.

Kamen was bothered by the shortcomings of the wheelchair. The wheelchair was supposed to make getting around easier for people with physical limitations, yet a curb posed a serious obstacle. Additionally, wheelchairs restricted people to a low sitting position, and many counters and cabinets were designed for people who could stand. Moreover, wheelchairs were likely to tip over. In the early 1990s, Kamen decided to create a more mobile, stable wheelchair.

In 2000 Kamen's improved wheelchair, the iBOT, which featured an elaborate balancing system, came onto the market. The system used gyroscopes to detect instability. When instability was detected, a computerized system in the chair repositioned the wheels to make the chair stable again.

A demonstraton in Tokyo of the Independence 3000 iBOT, developed by Kamen.

The iBOT balanced so well that it could operate on only two wheels. The chair had a four-wheel base that could tip upright so that only two wheels were touching the ground; this ability also made the chair taller, to allow a user to reach high counters and interact more easily with people who were standing. The base could also flip over, enabling it to climb stairs.

FROM FRED TO GINGER

Kamen sold the rights to the iBOT to Johnson & Johnson, but remained convinced that his balancing technology could be adapted for a day-to-day transportation device. He reserved the right to use the system in nonmedical devices. In 1998, DEKA's staff went to work on the new transportation device.

The new device was called Ginger, after Fred Astaire's dance partner, Ginger Rogers (the iBOT had been nicknamed Fred after Astaire, a famous dancer and movie actor). Ginger also used the iBOT's sophisticated balancing technology, but needed only two wheels. A rider

TIME LINE

1951	1971	1975	1982	1987
Kamen born; grows up in Rockville Centre, Long Island.	Kamen builds the first drug infusion pump.	Kamen founds AutoSyringe, Inc.	Kamen sells AutoSyringe to Baxter International.	Kamen's new company, DEKA, works on improved dialysis machine.

would stand on a platform between the wheels, steering with a handlebar that extended up from the wheelbase. Ginger was very responsive to cues from a rider and could travel up to 12 miles per hour (19 km/h).

Kamen was so confident about the technology used in Ginger that he decided to handle the device's manufacturing himself, rather than selling the rights to another company. He asked a journalist, Steve Kemper, to chronicle the development of what he felt sure would be an invention of great historic significance.

Kamen needed outside funding to develop and manufacture Ginger on a large scale, so he began showing it to various investors and technology luminaries including Steve Jobs of Apple Computer and Jeffrey Bezos of Amazon.com.

Kamen, left, and Jeffrey Bezos, head of Amazon.com; Bezos helped to hype the Segway and made it available online through Amazon.

TIME LINE (continued)

1993	1998	2000	2001
Kamen's dialysis machine goes on the market.	DEKA begins work on a secret new transportation project.	Kamen's iBOT wheelchair becomes available.	Amid widespread speculation, Kamen introduces his new invention, the Segway.

INCREASING HYPE

Kemper wanted to write a book about Ginger, so his agent submitted proposals to various publishers. The proposal's description of Ginger was left intentionally mysterious to prevent other writers from stealing the idea for the book, but included praise from Jobs and Bezos.

In January 2001, the proposal was leaked to a Web site; its appearance on the Web led to a flurry of online speculation about the nature of Ginger. Fanned by Kamen's renown as an inventor, the speculation spread to the mainstream news media, creating a frenzy; it was even suggested that Ginger could be a hovercraft or a device powered by hydrogen.

Although Kamen released statements calling some of the speculation "beyond whimsical," he had overestimated his invention's appeal. Kamen claimed, and apparently genuinely believed, that Ginger would revolutionize American society, altering the way cities were built in much the way the automobile had done.

On December 3, 2001, Kamen unveiled Ginger—now called the Segway—on the television show *Good Morning America*. The response of one of the show's hosts, Diane Sawyer, reflected the disappointment felt by many: she said, "But that can't be it." The Segway became available in 2003 to consumers at a cost of almost $5,000. Although sales numbers have never been released, the Segway was not a popular success, and, as of 2006, the device remains a novelty.

The Segway had no chance of living up to some people's expectations, but hype was not the only problem. Kamen had developed a technology that was very interesting from an engineering point of view but not as interesting to consumers. Throughout its development, Kamen had insisted that the Segway not be called a scooter. To the average consumer, however, that is exactly what the Segway is; the fact that it has an advanced balancing and steering system makes very little difference

> You have teenagers thinking they're going to make millions as NBA stars when that's not realistic for even 1 percent of them. Becoming a scientist or engineer is.
>
> —Dean Kamen

to someone who does not want a scooter in the first place. Dialysis machines and wheelchairs have a captive market; a scooter is physically necessary to almost no one, so better ones are not desperately needed.

Although enthusiasm for the Segway appears to have died down, Kamen has benefited from the publicity he received during the run-up to its unveiling. The publicity generated interest in his other projects, among them the FIRST robotics competitions (see box, For Inspiration and Recognition of Science and Technology). In addition, he has been

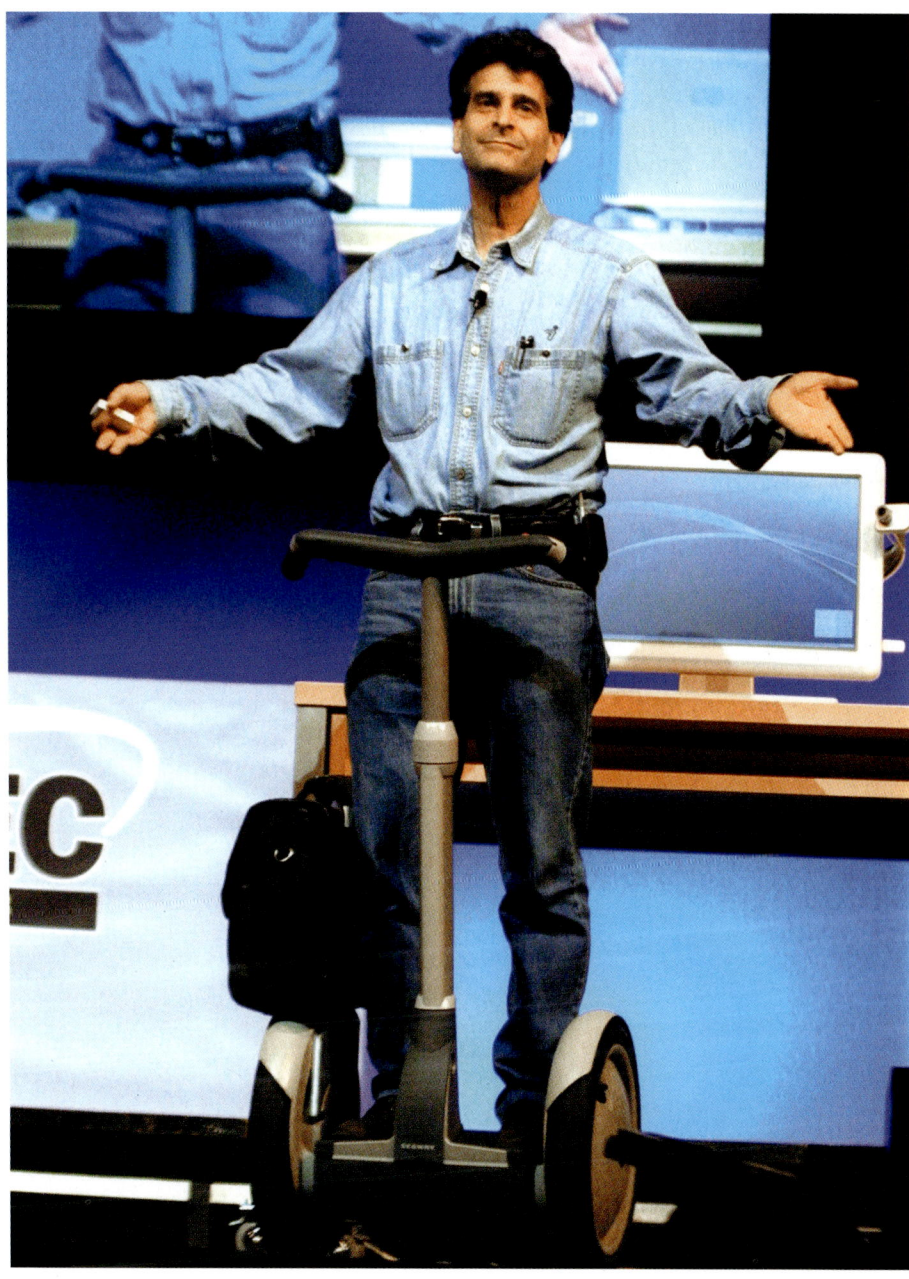

Kamen and his Segway human transporter in 2003.

developing an electrical generator that can be powered by wood or any other combustible material; such a generator could potentially bring electricity to poorer, rural parts of the world. For his work as a prolific inventor, Kamen received the National Medal of Technology in 2000 and the Lemelson-MIT Prize in 2002.

—Mary Sisson

Further Reading

Book
Kemper, Steve. *Code Name Ginger: The Story behind Segway and Dean Kamen's Quest to Invent a New World.* Boston: Harvard Business School Press, 2003.

Web sites
DEKA
   Web site of Kamen's research company.
   http://www.dekaresearch.com
FIRST
   Site of Kamen's robotics competition.
   http://www.usfirst.org
Independence iBOT 4000 Mobility System
   A Web site featuring information and videos about Kamen's improved wheelchair.
   http://www.independencenow.com
Segway
   Home of the company that manufactures Kamen's transportation device.
   http://www.segway.com

See also: Contests; Health and Medicine; Jobs, Steve, and Steve Wozniak.

NARINDER KAPANY

Inventor of fiber optics

1927–

Much of the information people exchange in the 21st century travels along fiber-optic cables—hair-thin strands of glass and plastic that carry messages coded inside pulses of light. Most telephone calls, televised images, and Internet data travel this way for at least part of their journey. Although fiber optics has a long history, it was named and popularized by Indian physicist Narinder Kapany in the mid-20th century.

EARLY YEARS

Narinder Singh Kapany was born in Moga, in the Punjab region of India, in 1927 and spent his childhood in the city of Dehra Dun. In school, he learned one of the most basic laws of physics: light always travels in a straight line. Kapany doubted this immediately: "It became almost an obsession of mine to bend light around corners." The idea stayed with him when he went to the university in the city of Agra; he experimented with prisms (wedge-shaped blocks of glass) trying to bend light, but it always seemed to travel in a straight line.

Kapany, ambitious and eager to make his mark on the world, was determined to find a way to bend light. In 1951 the perfect opportunity arose. He had just completed his bachelor of science degree at Agra University when an offer came to travel to London, England, to study for a higher degree in optics (the science of how light behaves). He enrolled at the Imperial College of Science and Technology, where his supervisor was Harold Hopkins (1918–1994), then a young physicist. In 1954, Hopkins and Kapany were approached by some English surgeons who wanted to develop a flexible gastroscope that could be inserted down patients' throats to see inside the stomach. The gastroscopes then in use were rigid and painful to swallow. The problem seemed to involve making light bend around corners down a tube—but how would that be possible if light always traveled in straight lines?

Narinder Kapany in a laboratory at Imperial College, London, in 1955.

EXPERIMENTS WITH LIGHT

Hopkins and Kapany were not the first to try to bend light. In the 1840s, Swiss professor Daniel Colladon found he could shine light down pipes filled with water. The light entered at one end and emerged at the other, even when the pipes bent around corners. Colladon realized that the water channeled the light in what he described as "one of the most beautiful, and most curious experiments that one can perform." Irish physicist John Tyndall achieved the same "light-pipe" effect in 1854. Today, Tyndall is widely known for discovering it, even though Colladon performed the same experiment and got the same effects a decade earlier.

Light pipes were being used in theater sets and grand municipal fountains by the end of the 19th century, but many years passed before other uses were found. In the 1920s, Scottish television pioneer John Logie Baird and his U.S. rivals C. Francis Jenkins and Clarence W. Hansell tried to use light pipes to transmit pictures. The following decade, German students Heinrich Lamm and Walter Gerlach tried to make a flexible light pipe that could be inserted into the throat to see inside the stomach. Although they had little success, they did manage to send a picture of the letter V down a very short piece of glass tubing.

INVENTING FIBER OPTICS

In London in the 1950s, Hopkins and Kapany wrestled with the problem of making a more flexible instrument. They were soon developing extremely thin glass fibers, around one-third the diameter of a human hair, and using them to transmit images. With many thousands of these fibers fastened together, they made a thick light pipe and sent images of the letters G-L-A-S down it.

TIME LINE

1927	1951	1954	1955	1960	1973	2000
Kapany born in Punjab, India.	Kapany goes to London to study optics.	Kapany and his adviser coin the term *fiber optics*.	Kapany finishes doctorate and begins teaching at the University of Rochester.	Kapany moves to San Francisco and founds Optics Technology.	Kapany starts second business, Kaptron.	Kapany founds third company, K2Optronics.

How Fiber Optics Work

". . . enough to set one dreaming."

The fiber-optic cables used in telephone lines are made from 100 or more optical fibers wrapped into a tight bundle. Each one of these fibers is a thin strand of glass or plastic that can carry millions of telephone calls or packets of Internet data at the speed of light. Lasers (powerful generators of pure light) send signals into one end of a fiber-optic cable, the signals travel down the cable to their destination, and light-detecting electronic circuits receive the signals at the other end.

Light signals travel through the cable by repeatedly bouncing off the glass walls. Although light can behave like a beam of rays, it can also be thought of as a steady stream of very energetic particles, known as photons. Imagine shining a flashlight into a glass rod so that photons of light pour into it. If the flashlight is held at an angle, the photons rush into the glass and travel in a straight line until they hit the edge of the tube. Then, just like rubber balls hitting a wall, they bounce back at the opposite angle. They keep reverberating through the tube until they reach the other end.

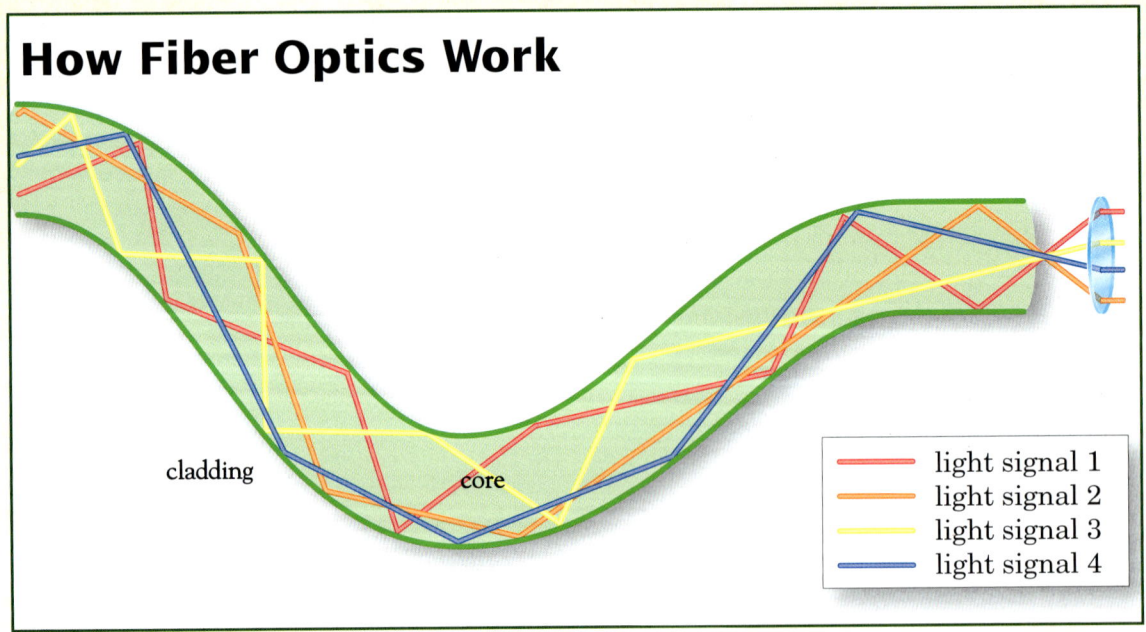

The process of total internal reflection: light travels through the core of the fiber and bounces off the cladding.

The key to fiber optics is making the light stay inside the tube. Normally, if light is shone into the end of a glass tube, some of it leaks out of the edges, with the result that it does not normally travel very far down a glass rod. An optical fiber is specially designed to stop leakage. The central section of the fiber (known as the core) is surrounded by a different type of glass (known as the cladding). Although the cladding is clear glass, it is designed so that it helps to "mirror" the light rays traveling through the core by a process known as total internal reflection.

The initial attempts made by Kapany and Hopkins transmitted light poorly. However, in the words of South African physician Basil Hirschowitz, "it was flexible, and did transmit an image, and that was enough to set one dreaming."

A hand runs a comb through the flexible glass fibers of a fiber-optic lamp; light is transmitted along the glass to the ends of the fiber.

Hopkins and Kapany were not the only scientists thinking about bending light. Abraham van Heel, a physicist working at the University of Delft in the Netherlands, was using similar technology to make submarine periscopes. By 1952, he had also figured out how to send images through bundles of fibers. He explained his ideas in an article for the scientific journal *Nature*; although the article was delayed several months, it was finally published in January 1954. Hopkins and Kapany, meanwhile, had sent details of their own results to *Nature*, and these were published in the same issue. Although the English team had used a different approach, dispute arose about who had thought of the idea first. Kapany and Hopkins called their idea "fiber optics," and the name stuck.

The following year, at the age of 28, Narinder Kapany earned his doctorate and decided to leave England for the United States. He accepted a position at the University of Rochester, where some of the best research in optics was being done. Later, he moved to the Illinois Institute of Technology in Chicago, where he became head of the optics department. During this time, he wrote several dozen important scientific papers that established the basic ideas of sending light down glass fiber-optic tubes. This work earned him the name "the father of fiber optics."

THE FIBER-OPTIC REVOLUTION

Kapany's work was still mostly theoretical; some obstacles had to be overcome before fiber optics became a useful technology. When many thin glass fibers were fastened together to make a thick light pipe, the light that leaked out through the sides of one fiber into the others tended to spoil any image traveling through the pipe. In the mid-1950s, Abraham van Heel coated optical fibers with an outer layer of cladding, a different kind of glass. Cladding traps more light inside the core (inner glass) of the fiber, so the light travels farther without leaking out.

Three scientists at the University of Michigan—Lawrence Curtiss, Basil Hirschowitz, and Wilbur Peters—used van Heel's invention to make another breakthrough. They solved the problem of seeing inside the body when they invented the fiber-optic gastroscope. This instrument was a bundle of thin optical fibers wrapped together inside a thick outer sheath that a physician could push down a patient's throat. Light shone down some fibers from a lamp outside into the patient's stomach. Some of the light reflected back up other fibers into the physician's eyepiece, showing an image of the stomach. Curtiss later said, "This was phenomenal. This was one of the few times in my life when I knew that I had something that was truly going to be significant." The fiber-optic gastroscope was used for the first time in 1957

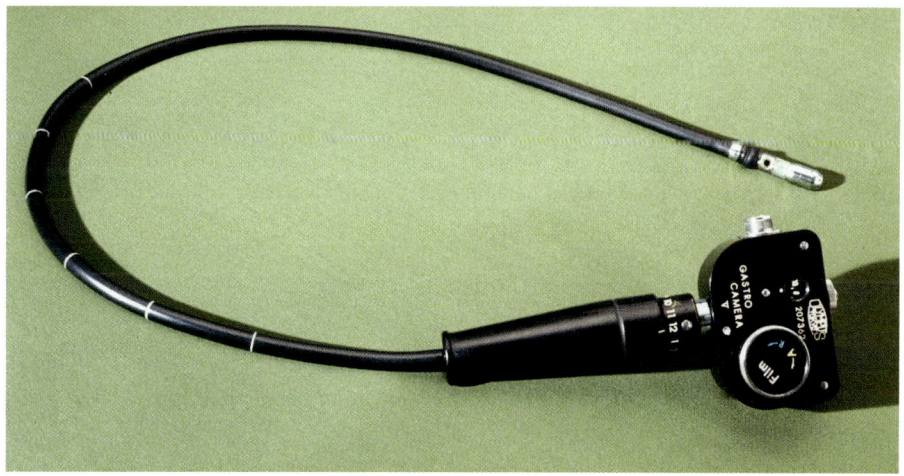

A gastroscope consists of a miniature camera with optical fibers that produce images of a patient's stomach.

and manufactured commercially in 1960; other medical instruments have since been invented that use fiber optics to see inside the body, thus saving many lives.

Medical optics was an important use for the new technology, but many people dreamed of something more: sending information down glass fibers at the speed of light. Even with cladding, early optical fibers could not carry light very far. The light's energy would fade, or "attenuate," as the light rays bounced repeatedly down the fiber. This was fine for gastroscopes, because they needed to carry light only a few feet or so, but it was not good for anything else.

A solution was devised in the mid-1960s by Chinese-born physicist Charles Kao (1933–), working in the United States at Standard Telecommunications Laboratories. Kao suggested that making fibers from very pure glass would allow beams of light to travel much farther. In 1970, a group of researchers at the Corning Glass Company achieved what Kao had described: a very thin, very pure optical fiber that could carry light signals long distances. Since then, fiber-optic cables have been widely used as telecommunications cables

KAPANY'S CONTRIBUTION

Like many modern scientists, Narinder Kapany capitalized on his research by going into business. In 1960, he moved to the Bay Area of San Francisco and started Optics Technology, a company that made fiber-optic parts for surgical instruments. He stayed with the company for 12 years before starting his second business, Kaptron, in 1973. In 2000, he and his son Rajinder founded a third company, K2Optronics, to develop advanced lasers for use with fiber-optic communication systems. For more than four decades, Kapany has balanced the roles of businessman and scientist, while continuing to patent new inventions and write books and scientific articles about the technology he helped to pioneer.

Narinder Kapany did not invent the fiber-optic gastroscope or telephone cable. He is known as the "father of fiber optics" because he was one of the key people who turned fiber optics from a science into a technology. Fiber optics began life as a 19th-century scientific discovery without a practical use. Following Kapany's work in the 1950s, people realized that this intriguing piece of science had practical uses, notably in medicine and telecommunications. Kapany was the scientist who helped to discover, name, and popularize the basic technology of fiber optics. Many others have built on his work, and today the industries based on fiber optics are worth many billions of dollars.

Kapany's companies have made him wealthy and have given him the chance to contribute to society in other ways, notably through his interest in the arts. In 1967, he established the Sikh Foundation, which helps to promote Sikh culture in the United States. Kapany is both a collector of Sikh art and an artist and sculptor in his own right, and his other passions include sailing, skiing, and participating in wildlife safaris in India. Although he has received many honors and awards, he remains modest about having helped to invent one of the most important communications technologies so far developed.

—Chris Woodford

Further Reading

Books
Arnold, Nick. *Frightening Light and Sounds Dreadful*. New York: Scholastic Hippo, 2001.
Burnie, David. *Eyewitness: Light*. New York: Dorling Kindersley, 1998.
Hecht, Jeff. *City of Light: The Story of Fiber Optics*. New York: Oxford University Press, 1999.

Web site
Optics for Kids
    Games, experiments, and lesson plans from the Optical Society of America. http://www.opticsforkids.com/

See also: Communications; Optics and Vision.

ANNA KEICHLINE

Inventor of the K brick
1889–1943

Anna Keichline was a female architect in a time when most buildings were designed by men. In her work, she made a point of designing rooms and appliances for ease of use. That approach led her to create a lightweight, inexpensive building material—the K brick—that eventually evolved into the hollow concrete blocks still used by builders today.

EARLY YEARS

Anna Wagner Keichline was born in Bellefonte, a town in Centre County, Pennsylvania, in 1889. Her father was a prominent lawyer, and the family was financially comfortable. As a child, Keichline was very interested in building. Her parents decided to encourage her interest and provided her with a complete carpenter's workshop.

In 1903, Keichline, then 14 years old, entered an oak card table and a walnut chest she had made in the Centre County fair. She won first prize, an instance considered so unusual given her gender and age that the *Philadelphia Inquirer* ran a story about her. The story noted that her pieces "in quality and finish, compare favorably with the work of a skilled mechanic," and said that Keichline "goes to school, but every spare moment is put in her shop."

Keichline graduated from high school in 1906 and attended Pennsylvania State University, studying mechanical engineering. She then decided that she wanted to study architecture and in 1907 transferred to Cornell University in New York.

Cornell had been built on land granted by the federal and state governments; as a result, the university was legally required to admit women. Nonetheless, at the time it was extremely rare for a woman to study or practice architecture, in part because architecture was so closely associated with the male-dominated fields of building and construction. When Keichline graduated from Cornell in 1911, she was only the fifth woman in its 46-year history to graduate with a degree in architecture.

BACK TO BELLEFONTE

Following graduation, Keichline returned to Bellefonte, where she began working as an architect, sharing an office with her father. She soon found work, and in the 1910s she designed schoolhouses, churches, apartment buildings, and homes throughout central Pennsylvania.

Keichline received her first patent in 1912 for a space-saving sink and washtub combination suitable for apartments. She would continue to patent inventions through the 1910s and 1920s,

Anna Keichline photographed around 1920.

A More Practical Kitchen

In 1924, Keichline filed a patent for a new design of a kitchen, "to improve the construction of the various objects in the kitchen for increasing the comfort of the housekeeper as well as reducing her work." Keichline believed that kitchens were designed very inefficiently.

Her designs had several features that can be found in many kitchens today but were far less common in the 1920s. The cabinet and stove fronts were built into the floor rather than standing on legs, to eliminate extra areas requiring cleaning. Additionally, the kitchen featured a motorized shaft that could be outfitted with various tools to peel potatoes, beat eggs, grind coffee, and so on—a precursor to the modern food processor. The shelves were made low enough to eliminate the need for a stepladder to reach them, and all cabinets had glass doors, allowing a cook to view their contents easily.

Keichline's improvements to the kitchen were a reflection of her belief that rooms should be designed by those who have experience using them. In Keichline's time, the majority of architects were men who were very seldom the dominant cooks in their households.

Illustrations of Keichline's more efficient kitchen design from 1924.

although she never made serious efforts to start businesses based on her ideas.

Keichline's work was interrupted by World War I, which the United States entered in 1917. She volunteered as a special agent in the military intelligence division of the army, working in Washington, D.C., until the war ended the following year.

Keichline returned to her work and to applying for patents. In 1920, a state licensing exam was made a requirement for all architects in Pennsylvania. Keichline passed the exam, officially becoming the state's first female architect.

THE K BRICK

Four years later, Keichline filed for a patent for a bed that could be folded into the wall. At this time, Keichline was very interested in improving the design of existing objects to make them easier to use. She felt that women were ill served by the male-dominated field of design: "Equipment of houses especially has been developed by people who seldom have experience using or operating these materials . . . and this can be done better by women than men. Indeed, it will never be accomplished until women take hold."

Keichline's focus on utility also led to her most influential invention, the K brick, which she patented in 1927. At the time bricks (usually made of fired clay) were made in the form of solid blocks. While solid bricks were strong, they were also heavy and used a lot of raw material.

In addition, builders often needed to use only a fragment of a brick to fit into a wall or into a particular design. Breaking bricks, however, was fairly difficult and tended to be imprecise: a solid brick could shat-

Keichline's K brick is easier for builders to use.

The Evolution of the Modern Brick

solid clay brick   K brick   modern brick

Many bricks used in modern construction have incorporated some of the innovations Keichline introduced with the K brick.

ter or break at the wrong point, wasting material and time. Having bricks custom-made for a particular use was expensive.

The K brick looked something like the Roman numeral two—II. A wall built of K bricks would look just like a solid brick wall on the outside, but be largely hollow inside. The K brick had many advantages over the standard brick: it used much less raw material, so it greatly lowered the cost of construction; it took less time to fire and thus could be made more quickly than other bricks.

A K brick had only about half the weight of a solid brick, much to the benefit of those who had to transport and place hundreds or thousands of bricks. The 10 notches built into the K brick allowed for easy and clean breaking into smaller pieces. Moreover, walls made of K bricks were hollow and thus could be easily filled with insulating or soundproofing material.

CONCRETE BLOCKS

The K brick immediately caught on with builders, and in 1931 Keichline was honored by the American Ceramic Society. She continued designing homes and buildings until her death in 1943.

Over the years, the K brick gave rise to several different shapes of hollow blocks, which are now more likely to be made of cement than clay. Hollow blocks, however, are still not as strong as solid blocks or poured cement. This disadvantage has been largely overcome with the development of reinforced concrete. Hollow-block walls can be reinforced with steel rods, which are placed vertically through the holes in the blocks and sealed into place with mortar, making them even stronger than a regular solid wall. Hollow blocks remain extremely popular among builders because they are easy and relatively inexpensive to use.

—Mary Sisson

TIME LINE

1889	1911	1912	1920	1927	1943
Anna Keichline born in Bellefonte, Pennsylvania.	Keichline graduates from Cornell University with a degree in architecture.	Keichline receives her first patent.	Keichline becomes the first female licensed architect in Pennsylvania.	Keichline patents the K brick.	Keichline dies.

Further Reading

Books

Kirkham, Pat, ed. *Women Designers in the USA: 1900–2000*. New Haven, CT: Yale University Press, 2000.

Macdonald, Anne L. *Feminine Ingenuity: Women and Invention in America*. New York: Ballantine, 1992.

Web sites

Anna Keichline
    An exhibit about Keichline by the Association of Women Industrial Designers.
    www.awidweb.com/pages/anna/anna_1.html

Anna Keichline
    A profile of the inventor from the Lemelson-MIT Program.
    http://web.mit.edu/INVENT/iow/keichline.html

See also: Buildings and Materials.

WILL KEITH KELLOGG

Inventor of cornflakes
1860–1951

When Will Kellogg was a child in the 1860s, the typical American breakfast consisted of heavy, hot foods, often pork or beef accompanied by some sort of starch. With so much fat and so little fiber in their diets, many people developed stomach disorders. Kellogg changed what many people eat for breakfast when he started making cereals from various grains, culminating in his greatest invention: the cornflake. Kellogg's Corn Flakes, with Will Kellogg's familiar red signature on the box, have been a staple of American households since the early 20th century.

EARLY YEARS

> Dollars do not produce character.
> —Will Kellogg

Will Keith Kellogg was born on April 7, 1860, in Battle Creek, Michigan, the town that would become famous for its breakfast cereal industry. Kellogg was written off as "dim-witted" by his teachers, but years later he would discover that he was merely nearsighted. He left school at age 15 and started working as a traveling salesman in his father's broom-making business.

His older brother, John Harvey Kellogg, was a charismatic physician and a devoted Seventh-day Adventist. Like most members of that health-oriented Christian denomination, he was also a vegetarian. John Harvey Kellogg became head physician and supervisor at the Battle Creek Sanitarium, a well-known health spa and hospital that had been founded in 1866 by the Seventh-day Adventists. One of his chief aims was to improve the health of his patients through a strictly vegetarian diet. When Will was 20 years old, he went to work as his brother's assistant.

GRANOSE

Will Kellogg became interested in nutrition while working at the sanitarium. Many of his brother John's ideas for dietary improvement were a little unusual, such as the all-grape diet, but others had more promise. The brothers began to research a variety of whole grain foods in order to provide patients with a healthy diet. They first developed a wheat flake, which all of the patients liked very much (see box, Inventing the Wheat Flake). The new flakes, called "Granose," became so popular that many patients asked to have packages sent to their homes after leaving the sanitarium.

Undated portrait of Will Keith Kellogg (artist unknown).

Will wanted to market their new product more widely. His older brother, however, insisted that the cereal be used only for his patients. The result was that one of the sanitarium's patients, C. W. Post, tasted the wheat flakes and set out to reproduce them. He succeeded—to the point that his own cereal company, which would produce such items as Post Toasties, would eventually become the General Foods Corporation. Other cereal makers in Battle Creek followed Post's example and began making similar wheat flakes; between 1900 and 1905, more than 40 cereal companies had begun operating there.

Will Kellogg was unhappy with his brother's lack of ambition, so when he succeeded in developing a similar kind of flake using corn instead of wheat, he determined not to miss out on the opportunity this time.

THE CORNFLAKE AND THE RISE OF KELLOGG

After much experimenting, he came up with what he felt was a better, crunchier, tastier flake. The brothers established Sanitas Food Company in 1898 to sell cornflakes by mail order. Although John was content to produce flakes on a small scale, serving only the sanitarium and its patients, Will had much bigger plans.

Intent on turning his cornflake business into a global packaged food enterprise, in 1906 Will Kellogg established the Battle Creek Toasted Flakes Company. Kellogg aggressively promoted his company and its product, and success came quickly: in its first year, the company shipped

Inventing the Wheat Flake

The cornflake has become the signature product of the W. K. Kellogg Company, but first came the wheat flake. The invention of the wheat flake came about one day by accident in the sanitarium where Will was assisting his brother, John Harvey Kellogg.

In an effort to create a more easily digested bread substitute for patients at the Battle Creek Sanitarium, the brothers were experimenting with boiling different grains prior to baking. One evening in 1894 or 1895, without meaning to, Will had let a pot of boiled wheat stand several hours before baking it. In the process, though Kellogg did not know this, the wheat had softened somewhat, or had become "tempered." Kellogg went ahead and rolled the wheat out and let it dry. He saw that each individual grain of wheat emerged within the mash as a thin flake. After he baked these flakes, he discovered that he had created a cereal that was very tasty, crisp, and easily eaten and digested.

The brothers served these wheat flakes, which they dubbed "Granose," to patients in the sanitarium.

175,000 cases of cornflakes. Within five years, Kellogg's Corn Flakes had found their way into the majority of kitchens across America.

With cornflakes as his staple product, Kellogg soon expanded his product line, introducing Kellogg's Bran Flakes in 1915, Kellogg's All-Bran in 1916, and Kellogg's Rice Krispies in 1928. In 1922 the business was renamed the W. K. Kellogg Company. Shortly thereafter, the company began operating in Canada, Australia, Europe, and Asia. Kellogg was one of the first American entrepreneurs to recognize the vast potential of international markets. He is also credited with the invention of such marketing techniques as including product giveaways and color advertising in magazines.

LATER YEARS

An advertisement for Kellogg's Corn Flakes from 1929 associates the cereal with the wholesome Camp Fire Girls.

Will Kellogg retired in 1929 as president of the W. K. Kellogg Company. He remained as chairman of the board until 1946. During this time he became increasingly involved with philanthropic activities. As early as 1925, he had formed the Fellowship Corporation to foster agricultural training. In 1930 he established the W. K. Kellogg Child Welfare Foundation after having been named a delegate to the White House Conference on Child Health and Protection by President Herbert Hoover. The Child Welfare Foundation then became the W. K. Kellogg Foundation; it remains one of the leading charitable institutions in the United States, donating more than $4.5 billion dollars between 1930 and 2006. The foundation has continued to focus on children's welfare; Kellogg strongly supported educating children and giving them the means to achieve independence and security because he believed the future of humanity depended upon it.

Kellogg spent his last years living mostly in California. He owned a horse ranch in Pomona and left this property to California State Polytechnic College for use as a campus. He was opposed to leaving his wealth to his children for fear that doing so would stifle their own ambition and independence. Kellogg died in Battle Creek on October 6, 1951.

Boxes of Kellogg's Corn Flakes on a conveyor belt at the Kellogg factory in Battle Creek, Michigan, in 1990.

KELLOGG'S IMPACT

Will Kellogg's legacy is twofold. First, his creation of the wheat flake and then the cornflake transformed the way people in America and all around the world start the day. The cereal industry that grew up in Battle Creek began with Kellogg's experiments at the sanitarium and soon expanded into the W. K. Kellogg Company. Today packaged breakfast foods can be found in nearly every kitchen in America and many other parts of the world. Over the years Kellogg expanded its offerings to include other convenience foods such as crackers, cookies, and meat substitutes. Among the company's many brands are Keebler, Pop-Tarts, Cheez-It, Morningstar Farms, Famous Amos, Chips Deluxe, and Eggo. The company's trademarked figures Tony the Tiger, Ernie Keebler, and others are among the most recognized characters in advertising.

TIME LINE

1860	1875	1880	1898	1906
Will Keith Kellogg born in Battle Creek, Michigan.	Kellogg leaves school to work as a traveling salesman.	Kellogg works as an assistant to his brother John.	The Kellogg brothers establish the Sanitas Food Company to sell their new cornflakes.	Will Kellogg establishes the Battle Creek Toasted Flakes Company.

TIME LINE (continued)

1922	1915–1928	1946	1951
Kellogg renames his company the W. K. Kellogg Company.	Bran Flakes, All-Bran, and Rice Krispies are introduced to the market.	Kellogg retires as chairman of the board of the Kellogg Company.	Kellogg dies.

Second, his philanthropic commitment led to one of the largest, most far-reaching charitable institutions in the world, the W. K. Kellogg Foundation. With headquarters in Battle Creek, the foundation makes grants to programs and projects relating to health, agriculture, and education in an effort to help people around the world gain independence. Will Kellogg was in the vanguard of early-20th-century entrepreneurs who embraced philanthropy as a way to address society's ills. His foundation served as a model for the great number of private charitable institutions that would appear across the United States in the economic boom years following the end of World War II in 1945.

—Paul Schellinger

Further Reading

Book
Carson, Gerald. *Cornflake Crusade*. New York: Rinehart, 1957.

Web site
W. K. Kellogg Foundation
   Biography of W. K. Kellogg.
   http://www.wkkf.org/WhoWeAre/Founder.aspx

See also: Food and Agriculture.

CHARLES KETTERING

Inventor of the self-starter for automobiles
1876–1958

Charles Kettering was an engineer and also one of the most famous inventors of his day. Whereas some inventors have grand ideas, Kettering focused on improving existing products to make them more practical. Despite having a seemingly limited focus, Kettering's inventions had an enormous impact on modern society. Among his many accomplishments was his help in transforming the automobile from a curiosity into a ubiquitous form of transportation.

EARLY YEARS

Charles Kettering was born August 29, 1876, on his family's farm near Loudonville, a small town in central Ohio. Kettering showed an early interest in machinery and won a scholarship to attend high school in Loudonville.

Kettering's family did not have the resources to send Kettering to college. After graduating from high school in 1895, he had to go to work. Kettering worked as a teacher for a year, then enrolled for summer classes at the College of Wooster, located in nearby Wooster, Ohio. The classes required a great deal of reading, and Kettering's eyes, which had never been strong, began to fail, forcing him to leave school and spend several months resting at home.

By the winter of 1897, Kettering's eyes had recovered enough that he could return to teaching. After a year and a half as a teacher, Kettering decided to again attend college. While at Wooster, he had

Charles Kettering at work in 1950.

seen a course catalog for Ohio State University in Columbus and discovered that the college had classes more to his liking.

In 1898, Kettering entered Ohio State to study electrical engineering. His eyes continued to trouble him, however, and he began to develop extremely painful headaches. After a year at Ohio State, Kettering was forced to drop out of college.

A despondent Kettering returned home. One day, while visiting a friend, he signed on to work for a telephone company digging holes for posts. He worked for the telephone company for almost two years, eventually advancing to foreman.

His health and vision seemed to improve with the outdoor labor, and, in 1901, Kettering reentered Ohio State. His eyes were still weak, but his classmates read books to him, and the university exempted him from a drawing class. Kettering was finally able to graduate with a degree in electrical engineering in 1904.

NATIONAL CASH REGISTER

While attending Ohio State, Kettering had continued to work for the telephone company, where he expected to stay full-time after graduation. Then a professor presented him with a promising opportunity: a job with National Cash Register Company (NCR) in Dayton, Ohio.

Cash registers of the time were mechanical devices powered by hand cranks. The keys were hard to depress, and the cranks tough to raise. NCR had done little to improve its product over the years, and sales were slowing. The company decided to hire Kettering to spearhead the effort to develop better products.

Kettering began work on developing department-store credit systems. For a customer to buy something on credit at that time, a clerk would have to call the store's credit department, which would look up the customer's record and either

Illustrations from Kettering's 1910 patent for an "electric driving device for registering machines," just one of a number of his cash-register patents.

approve or deny the purchase. If the purchase was approved, someone from the credit department would walk over to the clerk to deliver a slip indicating the approval. Kettering used his knowledge of telephone technology to design a system in which the credit department could remotely activate a stamping machine located at the clerk's counter that would print out the approval slip. The credit system went on the market in late 1904 and was a huge success.

Kettering was then assigned to another major NCR project: creating a cash register powered by an electric motor. Kettering designed an electric register by the end of 1905. Many small towns, however, did not have electricity. In 1908 Kettering designed a cash register that was powered by a spring, which was wound whenever the user closed the cash drawer. The next year, he designed a machine to be used in bank accounting; the machine was unusual because it could subtract as well as add.

MOONLIGHTING

In September 1909, Kettering suddenly resigned from NCR to focus on a side project he had been working on since mid-1908—a new electrical ignition system for automobiles.

At the time, automobiles were not very reliable. One serious weakness was the ignition system, which used an electrical spark to ignite the mixture of air and fuel in the engine's chambers to drive the car forward. The ignition systems created electricity by using small generators or batteries. The generator systems, however, tended to stall at low speeds, and the battery systems created a constant shower of sparks, running the batteries down very quickly.

> My interest is in the future because I am going to spend the rest of my life there.
>
> —Charles Kettering

Kettering was intrigued by the challenge, and he and some of his NCR colleagues began working nights and weekends on a solution. Kettering developed a battery-powered ignition system that would create a large spark only at the moment it was needed to ignite the fuel-air mixture. The result was a much more reliable ignition system that also preserved battery life.

Kettering's ignition system came to the attention of the Cadillac Motor Car Company, a maker of luxury automobiles. In July 1909, Cadillac ordered 8,000 ignition systems, taking Kettering and his moonlighting colleagues by surprise. The men hastened to find a company that would manufacture the systems. Kettering patented his ignition system and the men decided to give themselves a name: Dayton Laboratories and Engineering Company, better known as Delco.

Before Kettering, cars had to be started by hand-crank; here, Maude Odell, the first woman cab driver in New York City, starts her car in 1923.

DELCO

For his next project, Kettering decided to develop an electric self-starter. At the time, to start a car, the driver had to stand at the front of the car and crank the engine manually. Cranking cars by hand was tiring and could be dangerous, because the crank could fly off or the car could move. A self-starter would allow a driver to start a car from behind the wheel with the push of a button or the turn of a key.

An electric starting motor, however, would have to exert a large amount of power to start the main engine. Designers at the time assumed that any motor so powerful would be too bulky to fit into a car. Kettering, however, had designed electric cash registers, which had also needed small but powerful motors. Kettering realized that, like his cash register, a starting motor had to generate only a short burst of power. Kettering eventually combined his starting motor with the ignition and lighting systems.

By late 1911, Kettering had largely perfected his electrical self-starter, and Cadillac ordered 12,000 units. Delco could not find another company to manufacture the complex system, so Kettering and his colleagues borrowed money to create their own manufacturing operation.

The Delco self-starter appeared in the 1912 model Cadillac and was an immediate hit. By the end of the year, Delco had contracts with several other automobile makers, and the company expanded to around twelve hundred employees.

From left, Kettering, Alfred P. Sloan, Jr., and Nicholas Dreystadt, president of Cadillac Motor Car, inspect the first self-starter.

GENERAL MOTORS

The rapid expansion of Delco created headaches for Kettering, who was not very interested in running a business. In 1916, he and Delco's other owners agreed to sell the company to what would become General Motors, which was buying up a variety of automobile businesses, including Cadillac. Kettering went on to start other businesses, including one that developed the first home generators to serve rural areas without power lines.

In 1920, General Motors decided that it needed a research division, and executives approached Kettering to head it. Seeing a chance to focus on research and invention, Kettering agreed. His work was already having a major impact on the automotive industry. In the five years following the introduction of the self-starter, the number of cars sold increased sevenfold. His partnership with General Motors would help make the company into the world's largest manufacturer of automobiles.

Kettering did not try to completely remake automotive technology. Instead, he developed incremental improvements designed to make cars more comfortable, easier to use, and more reliable. For example, Kettering and his staff helped reduce engine vibration and improved shock absorbers and brakes. They developed more durable paint, better lubricants, and shatterproof glass. They also worked to make automobile engines more powerful and more efficient. The result was a better vehicle—one that more people wanted to buy.

Knock, Knock

One of the serious problems designers of engines had to cope with was a phenomenon called knock, which was caused by small explosions of the fuel-air mixture inside engines. Knocks not only alarmed drivers but also could damage engines and limit their power. Knocks became worse if the mixture of air and fuel in the engine was tightly compressed before it was ignited—one way to make an engine more fuel-efficient and more powerful.

In the 1910s, engine design was assumed to be the cause of knock. Kettering, however, suspected that the composition of gasoline might be the culprit. Tests of different gasolines supported his hunch, and more tests revealed that adding certain chemicals to gasoline eliminated knock. These chemicals, however, also damaged engines. In the early 1920s, Kettering's staff developed a mixture that eliminated knock and did not hurt car engines.

The gasoline, branded Ethyl gasoline, went on sale in 1923, and eventually became the industry standard. The knock-free gasoline also made possible the design, by Kettering, of high-compression gasoline engines that used one-third less fuel than earlier engines.

But Ethyl gasoline had a problem: its primary knock-suppressing compound contained lead, which was poisonous. At first, health officials assumed that only the people who made Ethyl gasoline were at risk, but eventually they realized that the lead discharged in car exhaust posed a risk to everyone, especially children, and leaded gasoline was banned. Nonetheless, Kettering's insights into how gasoline affects engine performance were important, leading to the development of modern high-octane gas.

Kettering and his team developed higher-quality gasolines.

BEYOND THE AUTOMOBILE

Kettering led efforts to improve technologies outside the automobile industry. For example, General Motors purchased the troubled Frigidaire refrigerator company in 1918. At the time, refrigerators were cooled by water or by air, so they were bulky, expensive units that required a lot of upkeep.

Kettering wanted the refrigerator to be a one-piece unit that a home owner could simply plug into a wall. His staff began looking at chemical refrigerants, but the only ones on the market were poisonous or explosive. Finally, in the late 1920s, Kettering's staff discovered that a class of chemicals, chlorofluorocarbons, were effective refrigerants, did not corrode machinery, and were completely nontoxic. Decades later, chlorofluorocarbons were found to harm the earth's ozone layer, so they were replaced with other chemicals, but not before they had helped to make home refrigerators and air-conditioning systems safe and practical.

A Frigidaire refrigerator, made by a subsidiary of General Motors in the early 1920s.

TIME LINE

1876	1898	1904	1905	1909
Charles Kettering born in central Ohio.	Kettering enrolls at Ohio State.	Kettering graduates with a degree in electrical engineering; takes job with National Cash Register.	Kettering designs electric cash register.	Kettering resigns from NCR and founds Delco.

A PUBLIC FIGURE

Kettering became quite well-known during the Great Depression and was even featured on the cover of Time magazine on January 9, 1933.

During the Great Depression of the 1930s Kettering became a public figure, as he tirelessly espoused his belief in the potential for new technologies and products to spark an economic recovery. He became a popular speaker and writer, coining sayings such as, "If you have always done it that way, it is probably wrong," and "It doesn't matter if you try and try and try again, and fail. It does matter if you try and fail, and fail to try again."

Since Kettering believed that new technology would help the economy recover, he went to work on developing some. He became interested in improving diesel engines, which at the time were extremely large and heavy, and therefore often limited usefulness. Kettering came up with several improvements, creating a smaller, lightweight diesel engine. In the mid-1930s the first diesel locomotives that used Kettering's engine were put in service. Diesel engines would eventually replace steam engines on the railroads, and lightweight diesel engines would also become common in ships, submarines, and cars and trucks.

Kettering retired from General Motors in 1947. He remained active, however, consulting with General Motors and conducting investigations into the nature of photosynthesis. In November 1958, Kettering collapsed suddenly—he had suffered a stroke. He died a few days later.

Kettering was a tireless innovator who was fortunate enough to be in a position where his ideas were likely to be developed. His style of invention—making dozens of small improve-

TIME LINE (continued)

1912	1916	1920	1947	1958
Cadillac's new model features the Delco self-starter.	Kettering sells Delco to General Motors.	Kettering takes job as head of GM research division.	Kettering retires.	Kettering dies.

ments to an existing technology—proved well suited to the world of business, as did his focus on making technology more practical. As a result of his work, automobiles, refrigerators, air conditioners, high-quality gasoline, and diesel engines have all become widespread and popular.

—Mary Sisson

Further Reading

Books
Boyd, T. A. *Professional Amateur: The Biography of Charles Franklin Kettering.* New York: Dutton, 1957.

Leslie, Stuart W. *Boss Kettering.* New York: Columbia University Press, 1983.

Young, Rosamond McPherson. *Boss Ket: A Life of Charles F. Kettering.* New York: Longmans, Green, 1961.

Web sites
Charles F. Kettering
　A profile of the inventor from the Lemelson-MIT Program.
　http://web.mit.edu/invent/iow/kettering.html

Hall of Fame/Inventor Profile
　A sketch of Kettering from the National Inventors Hall of Fame Foundation.
　http://www.invent.org/hall_of_fame/86.html

People & Events: Charles F. Kettering, 1876–1958
　Information on the inventor from PBS.
　http://www.pbs.org/wgbh/amex/streamliners/peopleevents/p_kettering.html

See also: Carrier, Willis; Diesel, Rudolf; Ford, Henry; Transportation.

INDEX

Page numbers in **boldface** type refer to article titles.
Page numbers in *italic* type refer to illustrations or other graphics.

abacus, 804
Abbans, Claude Jouffroy d', 651
accidents and mistakes
    cellophane, 829–830
    microwave oven, 833
    penicillin, 783
    soy compounds, 908–909
    vulcanized rubber, 723
    wheat flakes, 939
advertising and marketing
    Apple products, 887, 889, 890
    breakfast cereals, 940, *940*, *941*
    electric products, 862–863
    safety razor, 687, 688, 689
    Singer sewing machine, 842
    women-targeted, 832, 834, 835, 836
Aerospace Corporation, 678, 680, 681, 682
Aiken, Howard, 822
air-conditioning, 828–829, 830, 835, 950, 952
aircraft, 802, 860, 862
    aluminum, 755
    jet, 804
air pump, 817
air purifier, 729, 735–736
Alcoa, 755–756
All-Bran, 940, 942
Altair 8800, 885–886
aluminum, 749–756
Amazon.com, 918
American Safety Razor Company, 686, 689
Ampex Electric and Manufacturing Corp., 692–697
Analytical Engine, 871–872
anatomy, 781, 784
ancient civilizations
    drugs, 780
    hypocaust, *827*, 828
    inventions, 797–798, 801, 803
    medicine, 779, 781
    smallpox outbreaks, 876

Anderson, Charles, 695–696
anesthetics, 780, 783, 785
animation, 889
antibiotics, 780, 783, 789
anti-inflammatory drugs, 780
antioxidants, 761
antiseptics, 780, 782, 785
Appert, Nicolas-François, 832, 833
Apple Computer, 883–892, 918
Archimedes of Syracuse, 798
Aristarchus, 666
Aristotle, 663–664, 665
Arkwright, Richard, 765, 768–770
Armstrong, Neil, 704
artificial limbs, 786
aspirin, 780, 786
astronomy
    Galileo, 665–672, 673, 801
    Huygens, 849, 851–852, 856, 857
atomic medicine, 784
atom splitting, 676
automobiles
    aluminum engine, 755
    diesel, 951, 952
    emissions controls, 812–813
    engine knock, 949
    GPS receivers, 681–682
    Honda, 811–814
    innovations, 863–864
    Kettering improvements, 943, 946–952
    painting system, 734
    self-starter, 943, 946–947, *948*, 952
    vulcanized rubber, 726, 728
AutoSyringe, 788, 915, 919

Babbage, Charles, 867, 871
bacteria, 782, 783, 786, 789
bacteriology, 782
Baird, John Logie, 925
balance springs, 817
Barberini, Maffeo, 671

Bardeen, John, 829, 836, 860
Barnard, Christiaan, 783, 788
barometer, 817–818
BASIC (computer language), 886–887
Bath, Patricia, 783, 789
bathroom inventions, 830, 833–834
Battle Creek Toasted Flakes Company, 939–940, 941
Baxter International, 788, 915, 916, 919
BBC network, 712
Becquerel, Antoine-Henri, 785, 786
Bell, Alexander Graham, 799, 804
Benz, Karl, 863
Berger, Frank, 780, 788
Berners-Lee, Tim, 804, 889
Bethlem Royal Hospital (Bedlam), 819–820
Bezos, Jeffrey, 918, 920
Bible, printed, 741, 742, 743, 745, 746, 747
bicycles
    small engines, 805, 807, 811
    tires, 726
    See also motorcycle
Birdseye, Clarence, 832, 835
Blanchard, Helen, 843
block printing, 739, *741*
blood banks, 783, 787, 789
Blount, Bessie, 787, 788
blue jeans, 861
books, 737, 739, 746, 800, 863
Boyer, Herbert, 786, 788
Boyle, Robert, 817, 820
Brandenberger, Jacques, 829–830, 835
Bran Flakes, 940, 942
Brattain, Walter, 829, 836, 860
Braun, Wernher von, 707–708
breakfast cereals, 937, 938–942
Brown, Louise, 788, 789
Bruno, Giordano, 667

INDEX   953

buildings and materials
    aluminum, 751–756
    K brick, 931, 934–936
    London rebuilding, 819
    practical kitchen, 933
burner, automatic, 902, 903

Cadillac Motor Car Company, 946, 947, 948
Calabar beans, 908
calculator, pocket, 885
canals, 646–648, 651, 654
cancer treatment, 785
canning, 832, 833
    vacuum, 899–904
carbon paper, 794
Carlson, Chester, 860
Carothers, Wallace, 722, 728, 787, 803, 804, 833, 835, 860
Carrier, Willis, 829, 830, 835
cash register, 794, 945, 946, 947, 951
Cassini-Huygens Program, *851*, 856–858
cataract surgery, 783, 789
CBS, 694, 697, 711, 714, 715–717, 718
CDs (compact discs), 717
cell, coining of term, 818
cellophane, 829–830, 835
cell phones
    design, 889
    GPS receivers, 682
Census Bureau, U.S., *871*, 872
China
    gunpowder, 798, 802
    herbal remedies, 780
    magnetic compass, 799, 802
    printing, 739–740, *741*, 800
    rocket invention, 707
    smallpox immunity, 876, 877
    toilet paper, 834
chlorofluorocarbons 830, 950
cholera, 659, 782, 786, 788
chronometer, 773–778
circuit, flip-flop, 676
Civic compact (Honda car), 812–813
Clermont (steamboat), 652–653, 655

clocks
    chronometer, 773–778
    pendulum, 674–675, 774, 775, 776, 849, 852–853, *854*, 855, 856
cloning, 786, 788
cloth and apparel
    jacquard pattern, 865, 867–870, *871*, 872
    manufacture, 765–772, 801–802, 803
    sewing machine, 837–844, 846–847, 862
    waterproof, 720
COBOL (computer language), 824, 825
Cochran, Josephine, 830, 832
Cohen, Stanley, 786, 788
Cold War, 831
Cole, Romaine, 754
Colladon, Daniel, 925
color television, 711–718, 830, 831
compass, 799, 802
computers
    Apple, 883–892
    basic elements of PCs, 888
    compiler, 821–826
    debugging, 823
    history, 803, 804, 860
    innovation, 863, 864
    mouse, 829, 830
    punch cards, 871–872
    temporary memory, 676
    user-friendly systems, 829, 836, 887, 890–891
    wireless, 803
    See also Internet
concrete, 935
contests, 917, 921
Cook, James, 777–778
cooking, 828, 834, 836
coolants, 829, 830, 832
Cooley, LeRoy, 901, 902, 903
Copernicus, Nicolaus, 666, 667
Cormack, Allan, 784, 789
cornflakes, 937–941
corporate inventions, 804, 834, 859, 860, 863
cortisone, 909, 910
cosmic rays, 676
cottage industry, 769–772

cotton gin, 801, 802, 803, 844
cotton textiles, 764–767, 770–772
cowpox, 874, 878–879, 881
crane, 798
credit systems, 945–946
Cremoni, Cesare, 668
Crick, Francis, 785, 788
Crompton, Samuel, *765*, 770
cryolite, 752
crystals, 818
CT scanner, 784, 789
cuckoo nesting habits, 874, 875
Cummings, Alexander, 830, 833–834
Curie, Marie, 785, 786
Curtiss, Lawrence, 928
Curtiss-Wright, 707
CVCC Honda engine, 812–813, 814
cytotoxic drugs, 780

Daimler, Gottlieb, 863
Damadian, Raymond, 785, 789
Davies, Jacob, 861
DEKA Research and Development Corp., 915–916, 918, 919, 920
Delco, 946–948, 951, 952
de Mestral, George, 799
denim, 861
Descartes, René, 850
diabetes, 786, 788
dialysis machine, mobile, 788, 916, 919, 921
Diamond Sutra, *738*, 739
diesel engine, 951, 952
Digital Equipment Corporation, 824–825
dishwasher, 830, 832, 834
disposable products, 689, 833, 834, 860
DNA, 785–786, 788
Dolby, Ray Milton, 695
dolls, 793
"Dolly Dips" (sponge), 793
Drew, Charles, 783, 787
drinking water, 657–662
drugs, 780, 786, 788
    infusion pump, 788, 915, 919
    synthetic, 905, 907–911

dry-powder paint, 729, 733, 734
DuPont Company, 722, 803, 860
dynamics, science of, 856
Dyson, James, 832, 836

Eckert, J. Presper, 823
Eckert-Mauchly Computer Corporation, 823
Edison, Thomas
   electric power, 803, 804, 827, 828, 834, 863
   invention process, 799
educational toys, 793
Edwards, Robert, 788
elasticity, 728, 817
electric fan, 828, 834
electric guitar, 860
electricity
   car self-starter, 947
   household appliances, 827, 832, 834, 836
   ionized gas-generated, 731–732
   power plants, 803, 804, 827, 828, 834, 863
   product marketing, 862–863
electric motor, 833, 834
electric shaver, 688
electrodes, 752
electrogasdynamics, 729, 731–733, 734
electromagnetism, 784
electron, 752
elevator, 862
Eli Lilly Company, 780
energy and power
   alternate fuels, 812
   heat theory, 818, 828
   renewable, 659, 798, 801
   See also specific sources
Energy Innovations (company), 732
Engelbart, Douglas, 829, 830, 836
engines. *See specific types and uses*
environment and inventing
   air purifier, 729, 733, 735–736
   Gadgil projects, 657–662
   Honda emissions controls, 811–814
   ozone hole, 830, 950
Equanil (drug), 780
escalator, 862

escapement (gear), 673
Ethyl gasoline, 949
evolution theory, 818
EVR (electronic video recording), 717, 718

Fahrenheit, Daniel, 781, 784, 801, 803
falling objects, speed of, 663
fan, electric, 828, 834
Faraday, Michael, 833, 834, 862–863
Federal Communications Commission, 714, 715–716
feeding apparatus, 787, 788
fetal tissue, 786
fiber optics, 786, 787, 788, 923–930
filtration systems, 729, 733, 735–736
fire, 828
FIRST competition, 917, 921
Fisher, George, 840
Fitch, John, 651
flash-drying method, 759, 760
Fleming, Alexander, 780, 783, 786
flip-flop circuit, 676
fluoxetine (Prozac), 780, 788
flush toilet, 830, 833–834
flying shuttle, 767
folio, 743
food and agriculture, 797
   Kellogg products, 937–942
   kitchen inventions, 828, 829–833, 834, 836
   machinery, 801, 803
food preservation, 832, 833
   cellophane, 829–830, 835
   freezing, 832, 835
   Gadgil processes, 757–762
   pasteurization, 782, 785, 799
   plastic container, 728, 829, 830, 835
   vacuum canning, 899–904
Ford, Henry, 863
Franklin, Benjamin, 654, 828, 833
Franklin stove, 828, 833
freezer, 830, 832
French Revolution, 648, 867, 870
Frigidaire Company, 830, 950
frozen foods, 832, 835

Fujisawa, Takeo, 807, 808, 809, 811, 814
Fuller, George, 862
Fuller, Ray, 780, 788
Fulton, Robert, **645–656**, 803
Fust, Johann, 744

Gadgil, Ashok, **657–662**, 788–789
Galen, 779, 781, 784
Galilei, Galileo, **663–674**, 801, 802, 853, 855
gasoline engine, 801, 862
   CVCC Honda, 812–813
   knock problem, 949, 952
gastroscope, 788, 924, 928–929, 930
Gayetty, Joseph, 834
gears, 798, 853
General Electric, 832, 834
General Foods, 939
General Motors, 948–952
generator, 731–732, 733, 862–863, 922
genetic engineering, 786
genetics, 785–786, 788
geocentric universe, 666–667, 670, 671
Gerlach, Walter, 925
germ theory, 782, 785
Getting, Ivan I., **675–682**
Gibbs, James, 843
Gillette, King C., **683–690**, 833, 834, 860
Gillette Safety Razor Company, 686–687, 688, 689
Ginsburg, Charles, **691–698**, 829, 836
glaucoma, 780, 905, 907
   treatment, 908, 911
Glidden Company, 907, 909, 910
Global Positioning System, 675, 678–682, 774, 775
Goddard, Robert H., **699–710**
Goldmark, Peter, **711–718**, 830
Goodyear, Charles, **719–728**
Goodyear Tire & Rubber Company, 725, 726
Gourdine, Meredith, **729–736**
Gourdine Systems, 732, 735
Graham, Bette Nesmith, 829, 836, 860

gravity
: Galileo, 664, 665
: Hooke, 818
: Newton, 668–669, 672
Great Fire of 1666 (London), 818–820
Griffith, Carroll L., 758, 759
Griffith Laboratories, 758–761
gunpowder, 707, 798, 800, 802
Gutenberg, Johannes, **737–748**, 800, 802, 863
Gutenberg Bible, 741, 742, 743, 745, 747
gyroscopes, 703, 705

hacker (computer term), 886
Hall, Charles, **749–756**
Hall, Lloyd A., **757–762**, 832, 835
Hall-Héroult process, 753
Hancock, Thomas, 720, 725
Hansell, Clarence W., 925
Hargreaves, James, **763–772**, 801, 803
harness, 799, 802
Harrison, John, **773–778**
Hasbro, Inc., 896, 897
Hayward, Nathaniel, 722
health and medicine, **779–790**
: fiber-optic instruments, 928–929, 930
: inventions, 788, 801, 915, 916–918, 919
: smallpox vaccine, 873–882
: *See also* drugs
hearing aid, 820
heart disease, 881
heart transplant, 783, 788
heating, 827–828, 833
heat theory, 818, 828
Heel, Abraham van, 928
heliocentric universe, 666, 667, 668, 671, 673, 801
Henderson, Shelby, 696
Henry, Beulah, **791–796**
herbal remedies, 780
Héroult, Paul, 753
Hewlett-Packard, 884, 886, 890
hi-fidelity stereo, 717–718
Hippocrates, 779, 781, 784
hip replacement, 786
Hirschowitz, Basil, 928

history of invention, **797–804**
: innovative ideas, 859–864
Hoe, Richard March, 747
Hoff, Ted (Marcian Edward), 832, 885
Hoffmann, Felix, 780, 786
Honda, Soichiro, **805–814**
Honda Motor Company, 807–814
Hooke, Robert, 781, 784, 801, 802, **815–820**
Hooke's Law, 817
Hopkins, Harold, 924, 925, 928
Hopper, Grace Murray, **821–826**
hormone synthesis, 908, 909
Hounsfield, Godfrey, 784, 789
household inventions, **827–836**
: air purifier, 735–736
: aluminum pots, 755
: sewing machine, 837–844
: utilitarian design, 933, 934
Howe, Elias, **837–844**, 847, 862
Human Genome Project, 786
Hunt, Alfred, 754
Hunt, Walter, 828, 839, 841, **845–848**
Hunter, John, 874
Huygens, Christian, **849–858**
hydrostatic balance, 663
hypocaust, 827, 828, 833

IBM, 860, 891
iBOT wheelchair, 788, *916*, 917–918, 920, 921
ignition system, 946
illuminated manuscripts, 739, 741
iMac (computer), 890
immunization. *See* vaccines
I. M. Singer and Company, 842
incinerator filtration, 733, 735
inclined plane, 647
industrial design, 889, 890, 892
Industrial Revolution, 747, 771–772, 801–802
ink
: printing, 742, 743
: typewriter, 860
innovation. *See* invention and innovation
insulin, 786, 788
Intel Corporation, 832, 885, 890

Internet, 803, 804, 829
: fiber-optic cables, 923, 925
intravenous units, 914–915
invention and innovation, **859–864**
: myths about, 799
: *See also* history of invention
in vitro fertilization, 788, 789
ionized gas, 731–732, 736
iPod, 889, 890, 891
Islamic world, 799
Isle of Man Tourist Trophy race, 808, 809
isochronicity, 853, 855
iTunes, 890
Ive, Jonathan, 889

Jacquard, Joseph-Marie, **865–872**
Jenkins, C. Francis, 925
Jenner, Edward, 782, 785, **873–882**
Jenney, William Le Baron, 862
jet aircraft, 804
Jet Propulsion Laboratory, 731–732, 893, 896, 897
Jobs, Steve, **883–892**, 918, 920
Johnson, Lonnie, **893–898**
Johnson & Johnson, 918
Johnson Research & Development, 893, 897
Jones, Amanda, **899–904**
Julian, Percy Lavon, 780, 787, **905–912**
Julian Laboratories, 909, 910
Jupiter (planet), 667, 670, 672, 673

Kamen, Bart, 914–915
Kamen, Dean, 788, **913–922**
Kao, Charles, 929
Kapany, Narinder, 787, 788, **923–930**
Kaptron, 925, 929
K brick, 931, 934–936
Keichline, Anna, **931–936**
Kellogg, John Harvey, 938, 939, 941
Kellogg, William Keith, **937–942**
Kellogg's Corn Flakes, 937, 939–940, 941, 942
Kemper, Steve, 918
Kettering, Charles, **943–952**

keyhole surgery, 783
Khrushchev, Nikita, 831
kidney dialysis, 788, 916, 919, 920, 921
kinetic energy, 731
"kitchen debate" (1959), 831
kitchen inventions, 828, 829–833
    efficiency designs, 933
knife sharpener, 847, 848
Knight, Margaret, 829, 830, 834
Koch, Robert, 782, 786
Krems, Balthasar, 839
K2Optronics, 925, 929
Kwolek, Stephanie, 728

Laënnec, René, 781, 785
Lamm, Heinrich, 925
Langley, Samuel P., 802
languages, computer, 823–824, 825
Larami Corporation, 896, 897
lasers
    fiber-optic cable, 926, 927
    surgery, 783, 787
latex, 733
Latho (sponge), 793, 794
latitude, 773, 776
leaded gasoline, 949, 952
lecithin, 907–908
Leeuwenhoek, Antoni van, 801
Legos, 917
Leonardo da Vinci, 781, 784, 799
    flying machine, 862
levers, 798
Levitt, Theodore, 861
Libri, Giulio, 668
life expectancy (1900–2000), 789
light
    bending, 924, 925, 928
    fiber optics, 923–930
    ultraviolet, 660–662, 789
    wave theory, 818, 849, 853–856
lightbulb, 799, 834
Lindbergh, Charles, 703, 706
linotype machine, 747
Lippershey, Hans, 667–668
Liquid Paper, 829, 836, 860
Lister, Joseph, 782, 785
literacy, 737, 739
Livingston, Robert, 650, 653, 655
locks, canal, 646–647

Lodge, Oliver, 863
longitude, 773, 774–777, 778
looms. See weaving
LPs (long-playing records), 711, 714–715, 717–718
Lucent Technologies, 860
Luddites, 801
Luther, Martin, 746

Macintosh, Charles, 720
Macintosh computer, 890, 891
mackintosh raincoat, 720
Madersperger, Josef, 839
Magini, Giovanni, 668
magnetic compass, 799, 802
magnetic resonance imaging. See MRI
magnetic tape recordings, 692–694
magnetism, 734, 735, 834
manufacturing, 765–772, 802–803, 863
Marconi, Guglielmo, 799, 863
Mark computers, 822, 823, 824
marketing. See advertising and marketing
Maskelyne, Nevil, 777
Massachusetts Institute of Technology, 676, 677–678, 681, 686
mass production, 802, 863
Mauchly, John, 823
Maxey, Alex, 696
meat preservation, 759, 760, 761, 762, 832, 835
medicine. See health and medicine
mediums (psychics), 899, 900, 903
Mellon family, 754
Mergenthaler, Ottmar, 747
metal
    aluminum, 749–756
    movable type, 741, 742, 743, 747
    paint application, 729, 733, 734
microbiology, 818
microchip, 832–883, 885, 886
Micrographia (Hooke), 818, 820
microprocessor. See microchip

microscopes, 782, 818
    compound, 781, 784, 789, 801, 802
    reflecting, 817, 818
Microsoft, 863, 864, 890
microwave oven, 832–833
Middle Ages, 798–799
Midgley, Thomas, 830
military and weaponry
    computers, 822–823
    Global Positioning System, 678, 680–681
    gunpowder, 798, 800, 802
    naval designs, 648–649, 652, 655–656
    radar, 677, 678
    rockets, 706–708
    rubber uses, 722
Miller, Patrick, 651, 655
Miltown (drug), 780
"Miss Illusion" doll, 793
momentum, 856
Montagu, Mary Wortley, 877
moon landing, 699, 704, 709, 831
moons
    Jupiter, 667, 670, 672, 673
    Saturn, 851, 852, 856
Morita, Akio, 829, 836
Morse, Samuel, 804
Morton, William, 783, 785
motor, electric, 833, 834
motorcycle, Honda, 805, 807, 808–811, 812, 813, 814
mouse (computer), 829, 830
movie animation, 889
MP3 player, 889
MRI (magnetic resonance imaging), 785, 789
myths about inventions, 799

NASA, 707, 709, 856, 893
National Cash Register Co., 794, 945–946, 951
Native Americans, 780, 875, 876
Nautilus (submarine), 648, 649, 652, 654
navigation
    chronometer, 773–778
    Global Positioning System, 675, 678–682
Navstar satellite, 680, 682

neoprene, 722
Nerf guns, 897
Newcomen, Thomas, 802, 803
newspapers, 746
Newton, Isaac, 668–669, 672, 849, 856, 860
    Hooke rivalry, 815, 818
NeXT (computer company), 889, 891
Nicholas Machine Works, 794, 795
Nickerson, William, 686–687, 689
Nixon, Richard M., 831
nuclear physics, 676–677
nylon, 803–804, 835, 860
    medical uses, 787
    toothbrush, 833

odometer, 820
oil, 861, 902
Optics Technology, 925, 929
Otis, Elisha G., 862
oxygen-gasoline mix, 813
ozone hole, 830, 950

paddle-wheel boats, 650–653, 655–656
paint, 729, 733, 734
paper, 743
paper bags, flat-bottomed, 829, 830, 834
PARC (Palo Alto Research Center), 861
Parkinson, Bradford, 678, 681
Parkinson's disease, 786
Pasteur, Louis, 782, 785, 799
pasteurization, 782, 785, 799
pathology, 786
Paul V, Pope, 671
Paul, Les, 860
pendulum, 664, 774, 775, 776
    Huygen's clock, 852–853, 854, 855, 856
penicillin, 780, 783, 787
Peters, Wilbur, 928
Pfost, Fred, 696
photocopier, 860, 861
photons, 926–927
photosynthesis, 951
physostigmine, 780, 787, 908
pickle machine, 720–721
Pikl, Joseph, 908

pins, 828, 845–848
Pittsburgh Reduction Company, 754, 755, 756
Pixar (graphics studio), 889
plastics, 787, 829–830, 833, 834
plow, 801
pocket calculator, 885
pocket watch, 775, 776–777, 778
pollution. *See* environment and inventing
polymerization, 833
Post, C. W., 938
Post Toasties, 939
prehistoric tools, 797
printing, 737–747, 800, 802, 863
Pristley, Joseph, 733
progesterone, 909
programming
    Apple II computer, 886–887
    computer-language compiler, 821, 823–826
    punch cards, 865, 867, 868, 869, 871, 872
prosthetics, 787
Protestant Reformation, 746
protograph, 793–794
Prozac (drug), 780, 788
Psalter, 744
Ptolemy, 666–667
pulleys, 798
pump, 664
punch cards, 865, 867, 868, 869, 871
PVC (polyvinylchloride), 833

racial discrimination, 906, 907, 910
radar, 677, 678, 833
Radcliffe, William, 767
Radiation (RadLab) Laboratory (MIT), 677, 681
radio
    inventors, 799, 863
    magnetic tape, 693
    transistor, 829
radioactivity, 785, 786
radium, 785, 786
railroads
    diesel, 951
    steam, 802, 804, 861
razors, 683–690, 860
    disposable, 833, 834

RCA, 693, 715–716, 718
reduction, chemical, 752
refrigerator, 830, 832, 834, 950, 952
Renaissance, 799–800
Reno, Jesse, 860
research laboratories
    energy and environment, 658, 661–662
    food preservation, 759–762
    MIT RadLab, 677
rheumatoid arthritis, 909
Rice Krispies, 940, 942
Roberts, Ed, 885
Robinson, Robert, 908
robotics, 917, 921
Roche (company), 780
rockets, 677, 699–710, 894
    air-powered toys, 897
Rollerblades, 861
Roman Catholic Church
    Galileo heresy, 667, 670–671, 672, 673
    printed books, 746
Röntgen, Wilhelm, 784, 785, 786
Rowley, Charles, 848
rubber, 719–728, 787
rubrics, 743

safety burner, 903, 902
safety pin, 828, 845–848
safety razor, 683–690, 833, 834, 860
Saint, Thomas, 838–839
Sanitas Food Company, 939, 941
satellites
    Global Positioning System, 675, 678–682, 774, 775
    photography, 717
    See also space program
Saturn (planet), 851–852, 856, 857, 858
scanning, 784–785
Schawlow, Arthur, 783, 787
science, technology, and mathematics
    FIRST contest, 917
    Galileo theories, 663–674
    history of, 798, 799, 800–801
    Hooke inventions, 815–820
    printed books on, 746
scientific revolution, 663, 800–801

Sculley, John, 884, 888–889, 890
Segway, 788, 913, 918–921
Selsted, Walter, 694
sewing machine, 837–844, 846–847, 862
sextant, 885
Shäffer, Peter, 744
shaving. *See* razors
ships
    chronometer, 773–778
    diesel engine, 951
    Global Positioning System, 680
    steam-powered, 645, 652–653, 655, 802
Shockley, Walter, 829, 836
Sholes, Christopher Latham, 793, 829, 834
Sikh Foundation, 930
silicon chip. *See* microchip
silk weaving, 865, 866, 867–872
Singer, Isaac Merrit, 841, 842, 843, 847
skin autograft, 787
skyscraper, 862
slave trade, 801
smallpox
    described, 875
    vaccine, 782, 785, 873–882
    world outbreaks, 876, 881
sodium crystal, 759, 760
solar energy, 659
Sony Corporation, 829
Sony Walkman, 829, 836
soybeans, 907–909
space program, 894
    Cassini-Huygens, 851, 856–858
    Cold War rivalry, 831
    Galileo probe, 673
    Global Positioning System, 678, 679, 680
    rockets, 699, 703–710
    satellite photography, 717
Spencer, Percy, 833
spice sterilization, 760–761, 832
spindle, 766
spinning jenny, 763, 765, 767–772, 803
spinning mule, 765, 770–771
spinning wheel, 766, 767
spiritualism, 899, 900, 901, 903

sponge, 793
Sputnik (space satellite), 831
steam-driven presses, 742
steam engine, 802, 803
steamships, 645, 650–656, 802, 803
stem-cell research, 787
Stephenson, George, 802, 804
Steptoe, Patrick, 788
Sternbach, Leo, 780
stethoscope, 781, 790
stoves, 828, 833
Strauss, Levi, 861
Sturgeon, William, 834
submarine, 648–649, 652, 654
sun. *See* heliocentric universe
SuperCub (motorcycle), 811
Super Soaker, 893–898
surgery, 782, 783, 785, 787, 789
synthetics, 803, 860
    drugs, 905, 907, 908–910, 911
    medical uses, 786, 787
    rubber, 722, 728
    See also nylon; plastics

tapestry, 865, 867
telegraph, 804
telephone, 799, 804
    fiber-optic cables, 923, 926, 930
telescope
    Galileo, 663, 667–670, 672, 673, 801, 802
    Huygens lens, 849, 851–852, 856
    reflecting, 817
    reflecting or refracting, 669
television
    fiber optics, 923, 925
    history, 712–714
    videocassete player, 717
    videotape, 691, 692–698, 829
    See also color television
temperature scale, 781, 801
textiles. *See* cloth and apparel
thermometer, 781, 784, 801, 803
Thimonnier, Barthélemy, 839
Thompson, Benjamin, 828
thread spinning, 766, 767, 770–771
3M (company), 860
tires, 726, 728
Titan (moon), 851, 852, 856–858

toilet paper, 834
toilets, 830, 833–834
tools, prehistoric, 797
toothbrush, nylon, 833
torpedo, 648–649, 652
Townes, Charles, 783, 787
Toyota Motor Corp., 806–807, 811
toys
    bathtub, 833
    Henry products, 793
    Super Soaker, 893–898
tranquilizers (drugs), 780, 788
transistor, 829, 836, 860
transportation
    history, 799
    Segway, 918–920
    waterways, 645–656
    See also specific types
tuberculosis, 782, 786
Tull, Jethro, 801, 803
Tupper, Earl, 829, 830, 835
Tupperware, 728, 829, 830, 835
Tyndall, John, 925
typewriter
    correction fluid, 829, 836, 860
    Henry inventions, 793–794, 795
    Shales invention, 829, 834

ultraviolet (UV) light, 660–662
umbrella, snap-on cloth cover, 792–793, 794
undulatory theory, 855
UNIVAC-1 computer, 823, 824
universal joint, 818
Urban VIII, Pope, 671
UV Waterworks system, 657, 658, 659, 660–661, 789

V-1 and V-2 rockets, 708
vaccines, 782, 785, 873–882
vacuum canning, 899–904
vacuum cleaner, 832, 834, 836
variolation, 876, 877–878, 879
vegetarianism, 938
Velcro, 799
vellum, 743
Vesalius, Andreas, 781, 784
videocassette recorder (VCR), 697, 717
videotape recorder (VTR), 691–698, 829, 836

Viking probes, 709, 710
vinyl records, 715
Virchow, Rudolf, 781, 786
viruses, 782, 875, 882
vulcanization, 719–728

Walkman, 829, 836
Wan Hu, 707
washing machine, 831, 834
watches
    balance springs, 817
    digital, 885
    GPS receivers, 682
    pocket, 775, 776–777, 778
water boiler, 828
waterborne disease, 659, 782, 786, 788
water frame, 765, 768–769, 770, 771
water guns, 893, 895, 896–899
waterproof cloth, 720
water purification, 657–661, 788–789
waterwheels, 798, 801
Watson, James, 785, 788
Watt, James, 651, 802

wave theory of light, 818, 849, 853–856
WearEver (aluminum pots), 755
weaving, 763, 769, 801
    flying shuttle, 767
    Jacquard loom, 865, 868–871
    steam-powered, 802
West, Benjamin, 646–648, 654
wheat flakes, 938–939, 941
wheel, invention of, 798, 801
wheelchair, 788, 916–918, 920, 921
Wheeler, Schuyler, 828, 834
Whitney, Eli, 801, 802, 803, 844
Wickham, Henry, 722
Wilson, Allen B., 843
windmills, 798, 801
Windows (operating system), 863, 864
wireless technology, 803
W. K. Kellogg Company, 939, 940, 941, 942
W. K. Kellogg Foundation, 940, 942
Woman's Canning and Preserving Co., 899, 901, 902, 903

women
    laborsaving devices, 832, 834, 835, 836
    patent holders, 791, 795
Woodward, Thomas, 847
wool industry, 764, 765
word processing, 829
World Health Organization, 788, 881
World War II
    computers, 822–823
    Honda factories, 806, 807
    rockets, 706–708
    synthetic rubber, 722
World Wide Web, 804, 889
Wozniak, Steve, **883–892**
Wren, Christopher, 819, 820
Wright, Orville, and Wilbur Wright, 755, 860, 862
writing, invention of, 797–798, 801

Xerox Corporation, 861
x-rays, 784, 785, 786, 790
    MRI vs., 785

Young, Thomas, 856

PHOTOGRAPHIC CREDITS

Cover photos (clockwise from left): LUGO/iStockphoto; AFP/Getty Images; China Foto Press/Getty Images; Noah K. Murray/Star Ledger/Corbis. **Corbis:** ASA-JPL-Space Science Institute/Handout/CNP 851; Bettman 687, 691, 694, 730, 750, 791, 792, 794, 835, 883, 884, 907, 938, 943, 947, 948, 950; Werner Forman 738; Historical Picture Archives 685; Kim Kulish 885; Philippe Lissac/Godong 803; Roger Ressmeyer 887; Vanni Archive 827; Peter Yates 937, 941; **Richard Diehl:** 692; **Early Television Museum:** 711, 717; **Getty Images:** Slim Aarons/Hulton Archive 806; Tim Boyle/Getty Images News 949; Stephen Chernin/Getty Images News 729, 735; John Chillingworth/Hulton Archive 783; Thomas Coex/AFP 680; Bruce Dale/National Geographic 923, 927; Jonathan Daniel/Getty Images Sport 696; Alfred Eisenstaedt/Time & Life Pictures 676; Thomas S. England/Time & Life Pictures 894; Kenneth Garrett/National Geographic 797; R. Gates/Hulton Archive 832; Paul Hawthorne/Getty Images Entertainment 683, 689; Hulton Archive/Hulton Archive 663, 665, 720, 770, 782, 850, 862; Imagno/Hulton Archive 819; Cynthia Johnson/Time & Life Pictures 821, 825; Doug Kanter/AFP 865, 872; Keystone/Hulton Archive 712, 809; Toshifumi Kitamura/AFP 859, 918; Jimin Lai/AFP 805, 808; Simon Maina/AFP 837, 844; Shah Marai/AFP 873, 881; Leonard McCombe/Time & Life Pictures 944; Joseph McKeown/Hulton Archive 924; David McNew/Getty Images News 880; Francis Miller/Time & Life Pictures 905, 911; Ralph Morse/Time & Life Pictures 716; Tim Mosenfelder/Getty Images Entertainment 891; NASA/Handout/Hulton Archive 673; Guang Niu/Getty Images News 741; Kazuhiro Nogi/AFP 745; Scott Olson/Getty Images News 762; Ruth Orkin/Hutlon Archive 693; Spencer Platt/Getty Images News 864; Bill Pugliano/Getty Images News 755; RNHRD NHS Trust/Stone 910; Eric Schaal/Time & Life Pictures 714; David E. Scherman/Time & Life Pictures 784; Howard Sochurek/Time & Life Pictures 831; Stock Montage/Hulton Archive 646, 664, 779, 874; Justin Sullivan/Getty Images News 812; Mario Tama/Getty Images News 914, 919; Three Lions/Hulton Archive 721, 849, 852, 877; Time & Life Pictures/Time & Life Pictures 672, 677, 840, 863, 951; Mario Villafuerte/Getty Images News 913, 921; Phil Walter/Getty Images Sport 915; **Griffin Laboratories:** 757, 758, 761; **iStockphoto:** Anyka 764; binabina 675, 681; blondiegirl 893, 897; bradwieland 828; clu 737, 746; dsteller 909; H2OWorks 899, 900; hidesy 833; jgdudash 781; Lammeyer 740; Lingbeek 645, 655; lucato 785; Madibwana 722; MortonPhotographic 733; **Jupiter Images:** 719, 724, 726, 744, 763, 771, 798, 845, 848, 853, 866, 878, 904; **Library of Congress:** 648, 649, 684, 732, 749, 754, 822, 823, 842, 940; **NASA:** 699, 700, 702, 706, 708, 709, 857; **Nancy Perkins:** 931, 932, 933, 934; **Science Museum/Science and Society Picture Library:** 647, 650, 667, 668, 671, 768, 773, 774, 777, 786, 800, 802, 815, 816, 817, 818, 838, 869, 870, 929; **United States Patent Office:** 679, 686, 695, 734, 753, 795, 839, 846, 895, 901, 916, 945

```
R
609 I62
v. 3

Inventors and inventions.
Walter ADU REF
05/08
```